The Media and the Tourist Imagination

Tourism Studies and Media Studies both pose key issues about how we perceive the world. They raise acute questions about how we relate local knowledge and immediate experience to wider global processes and both play a major role in creating our map of national and international cultures.

The Media and the Tourist Imagination adopts a multidisciplinary approach to explore the interactions between tourism and media practices within contemporary culture; in which the consumption of images has become increasingly significant. The contributions are divided between those written from media studies awareness, concerned with the way the media imagine travel and tourism; those written from the point of view of the study of tourism, which consider how tourism practices are affected or inflected by the media, and those that attempt a direct comparison between the practices of tourism and the media. A number of common themes and concerns arise with particular emphasis upon the image as the object of consumption.

While exploring the overlapping roles of tourism and the media, the collection is also concerned to mark out their different approaches to the structuring and organizing of experience and the way in which this leads to a dynamic interchange between them. Tourism and the Media are discussed as separate processes through which identity is constructed in relation to space and place.

David Crouch is Professor of Cultural Geography and Tourism, Director of the Culture, Lifestyle and Landscape research group at University of Derby, and **Rhona Jackson** and **Felix Thompson** are Lecturers in Film and Television Studies at the University of Derby.

Contemporary Geographies of Leisure, Tourism and Mobility

Series editor: Michael Hall is Professor at the Department of Tourism, University of Otago, New Zealand

The aim of this series is to explore and communicate the intersections and relationships between leisure, tourism and human mobility within the social sciences.

It will incorporate both traditional and new perspectives on leisure and tourism from contemporary geography, e.g. notions of identity, representation and culture, while also providing for perspectives from cognate areas such as anthropology, cultural studies, gastronomy and food studies, marketing, policy studies and political economy, regional and urban planning, and sociology, within the development of an integrated field of leisure and tourism studies.

Also, increasingly, tourism and leisure are regarded as steps in a continuum of human mobility. Inclusion of mobility in the series offers the prospect to examine the relationship between tourism and migration, the sojourner, educational travel, and second home and retirement travel phenomena.

The series comprises two strands:

Contemporary Geographies of Leisure, Tourism and Mobility aims to address the needs of students and academics, and the titles will be published in hardback and paperback. Titles include:

The Moralisation of Tourism
Sun, sand ... and saving the world?
Jim Butcher

The Ethics of Tourism Development
Mick Smith and Rosaleen Duffy

Tourism in the Caribbean
Trends, Development, Prospects
Edited by David Timothy Duval

Qualitative Research in Tourism
Ontologies, Epistemologies and Methodologies
Edited by Jenny Phillimore and Lisa Goodson

The Media and Tourist Imagination
Converging cultures
Edited by David Crouch, Rhona Jackson and Felix Thompson

Routledge Studies in Contemporary Geographies of Leisure, Tourism and Mobility is a forum for innovative new research intended for research students and academics, and the titles will be available in hardback only. Titles include:

The Media and the Tourist Imagination

Converging cultures

Edited by David Crouch, Rhona Jackson and Felix Thompson

Routledge
Taylor & Francis Group

LONDON AND NEW YORK

First published 2005
by Routledge
2 Park Square, Milton Park, Abingdon, Oxon OX14 4RN

Simultaneously published in the USA and Canada
by Routledge
270 Madison Ave, New York, NY 10016

Routledge is an imprint of the Taylor & Francis Group

© 2005 David Crouch, Rhona Jackson and Felix Thompson

Typeset in Times by Taylor & Francis Books
Printed and bound in Great Britain by
The Cromwell Press, Trowbridge, Wiltshire

British Library Cataloguing in Publication Data
A catalogue record for this book is available from the British Library

Library of Congress Cataloging in Publication Data
A catalog record for this book has been requested

ISBN 0–415–32625–7 (hbk)
ISBN 0–415–32626–5 (pbk)

Contents

Illustrations

Tables

Figures

Contributors

Claudia Bell is Senior Lecturer teaching Media and Cultural Studies at the University of Auckland. She has worked on tourism projects since the early 1980s, including visitor surveys and site evaluations for various organizations, including the Deptartment of Conservation and the Waitangi National Trust. She has published extensively on tourism issues (books, articles, technical reports), including work on town promotion in New Zealand, museums, displaying nations at Expos, landscape as tourist attraction; and on the tourists themselves as collectors of experiences to construct their own autobiographies. She is the co-author, with John Lyall, of *Inventing New Zealand: Putting Our Town on the Map* (HarperCollins, 1995).

Gary Best lectures in tourism and management at La Trobe University, Melbourne, Australia. His research interests are in heritage, distinctive cultural interactions in touristic contexts, commodification, and organizational culture. He has been a cultural tourist since he first set off for London in 1976. Other areas of interest include popular culture and automotive history.

Neil Campbell lectures in American Studies at the University of Derby. His recent publications are, as editor of *American Youth Cultures* (Edinburgh University Press/Routledge USA, 2004), author of *The Cultures of the American New West* (Edinburgh University Press/Fitzroy Dearborn, 2000), and co-author of *American Cultural Studies* (Routledge, 1997). He is currently writing a new book, *The Rhizomatic West* for the University of Nebraska Press (due 2006).

Sara Cohen is Director of the Institute of Popular Music at the University of Liverpool, and lectures in the School of Music. She is author of *Rock Culture in Liverpool: Popular Music in the Making* (Oxford University Press, 1991). Her research interests have involved specific projects on music and urban regeneration. She is a member of the editorial board of the journal *Popular Music* published by Cambridge University Press.

Nick Couldry is Senior Lecturer in Media and Communications at the LSE. His publications include *The Place of Media Power, Media Rituals: A*

Critical Approach (Routledge, 2000), edited collections *Contesting Media Power* (with James Curran) (Rowman and Littlefield, 2003) and *MediaSpace* (with Anna McCarthy) (Routledge, 2004) . He is currently engaged in a long-term project on Media and Citizenship.

David Crouch lectures in Cultural Geography and Tourism at the University of Derby, he is Visiting Professor of Geography at Swedish Universities of Karlstad and Kalmar. His recent edited books include *leisure/tourism geographies* (Routledge, 1999), *Visual Culture and Tourism* (Berg, 2003), and *Cultural Turns/Geographical Turns* (Longman, 2000). His book *The Allotment: its landscape and culture* (Faber, 1988) positioned a new approach to thinking cultural geography. He has written numerous research journal papers and book chapters on cultural and rural geography, ethnographic research methods, tourism, leisure, nature and landscape.

Solange Davin is a freelance researcher. She has published articles exploring the reception of medical dramas on television. Her book on the reception of the American medical drama ER is due to be published early 2005 in France. She is currently editing books on medical dramas and on literature and television, to be published in 2006.

Marcella Daye teaches modules on Tourism in Developing Nations and International Tourism, Sports Tourism and an Introduction to Tourism at Coventry University. She has previously worked in public relations and marketing at the Jamaica Tourist Board. Her main research interests are destination image, consumer behaviour in tourism and the relationship between tourism and the media.

K. J. Donnelly is a Lecturer in the department of Theatre, Film and Television at the University of Wales, Aberystwyth. He is author of *Pop Music in British Cinema* (BFI, 2001), editor of *Film Music* (Edinburgh University Press, 2001) and has just finished another book on film music for the BFI. His research interests embrace British and Irish cinema, music and a peripheral interest in geography, including having an article reprinted in Michael Dear and Steven Flusty, eds., *The Spaces of Postmodernity: Readings in Human Geography* (Blackwell, 2002).

David Dunn was a television programme maker for thirty years, mostly directing drama, soap opera and arts documentaries. He has subsequently taught media theory and video production at the Universities of Wolverhampton, Salford and Paisley. His research interests include televisual representations of Scotland, and he has published on the Gaelic Soap Opera *Machair*, and *Castaway 2000*'s engagement with Hebridean culture.

Tim Edensor lectures in Geography at Manchester Metropolitan University. He is the author of *Tourists at the Taj* (Routledge, 2003), *National*

Identity, Popular Culture and Everyday Life (Berg, 2002), and most recently, *Industrial Ruins: Aesthetics, Materiality and Memory* (Berg, 2005). He has written widely on tourism, walking in the city and the countryside, the film *Braveheart*, and car cultures.

Robert Fish lectures in Human Geography at the University of Exeter. His research interest is in rural visual culture. He is editor of a forthcoming collection entitled *Cinematic Countrysides* (Manchester University Press) and is currently researching the role of political and commercial film making in processes of agricultural change during the early post-war period.

Richard Grassick was a founding member of the Film Co-operative Amber Films in the 1970s and his recent work at Amber includes documentary dramas for UK Channel Four TV and critical documentary photo-essays of contemporary cultural change, focused in the North-East of England, Germany and the Czech Republic. These have taken the form of exhibitions touring these regions in a mix of local shows and formal gallery spaces with documentary of changing lives of coal, shipyard and textile workers, and more recently featuring the exhibition 'Post Industrial'.

Rhona Jackson teaches a module, Television and Tourism on the Film and Television Studies and Media Studies Programmes at the University of Derby. Her research interests are television audiences and film spectatorship, celebrity and fandom, popular culture, and cultural identity. On the radio she has given a talk on 'Women, Television Fiction, and Cultural Life' (Radio Sheffield) and participated in a debate on 'Confessional Narratives: A Debate'; (ABC National Radio. Melbourne, Australia).

John Lyall is an artist and Head of Sculpture at UNITEC in Auckland, New Zealand, who exhibits in the genres of installation, photography, performance, and sound. His cyber-opera 'Requiem for Electronic Moa' was performed in Soundculture, Auckland, in 1999 and at the Nuffield Theatre, Lancaster UK, in 2000. His most recent photography exhibition was 'Transit of Auckland,' at the Lopdell Gallery, Titirangi, 1999–2000. He is the co-author, with Claudia Bell, of *Inventing New Zealand: Putting Our Town on the Map* (HarperCollins, 1995).

Phil Powrie is Professor of French Cultural Studies and Director of the Centre for Research into Film and Media (CRIFAM) at the University of Newcastle upon Tyne. He has published widely in French cinema studies, including *French Cinema in the 1980s: Nostalgia and the Crisis of Masculinity* (Oxford University Press, 1997); as editor, *Contemporary French Cinema: Continuity and Difference* (Oxford University Press, 1999); *Jean-Jacques Beineix* (Manchester University Press, 2001); *French Cinema: An Introduction* (co-authored with Keith Reader, Arnold, 2002). He is the general co-editor of the journal *Studies in French Cinema*, and is currently editing *24 Frames: French Cinema* (Wallflower Press), co-editing a volume on the films of Luc Besson (Manchester University Press), and co-authoring a monograph on film adaptations of the Carmen story.

Felix Thompson teaches Film and Television Studies at the University of Derby. He has published on the areas of Third World and Experimental Cinemas. His current research interests include World Cinema with a specific interest in Japanese Cinema and the role of film theory in the understanding of British Cinema.

1 Introduction

The media and the tourist imagination

*David Crouch, Rhona Jackson
and Felix Thompson*

In addressing the connections between the media and tourism this collection breaks new ground. There are a multitude of tourist practices and an extended range of media. We are, therefore, immediately engaged in a process of multiplication. There are many connections, overlaps and disjunctures between tourism and the media and equally between the disciplines of tourist and media studies which the authors here explore in diverse fashion. This new area of investigation throws up pluralism of debate from the very start. It is worth immediately emphasizing the fact that we are dealing with two disciplines as well as the two different objects of study. It will readily become apparent when reading the contributions that there is no one theoretical perspective, no single angle of approach and indeed no obvious starting point.

Nevertheless it is well to think of strategies to encompass such variety of discussion and offer potential for the assorted contributions in this new field of study to become productive through cross-fertilization. At one level this can be addressed by considering recurrent debates across the range of these contributions. At another level it is suggested here that there is an overarching and necessary interdependence between tourism and the media. This is explored through the notion of the tourist imagination. To discuss the tourist imagination as a kind of bridging concept is to recognize the shared vitality which lies as much in the sense of global mobility engendered by the media in our daily consumption of films, books, television, newspapers and photography as it does in the actual activities of travelling, enjoying and exploring. The media are heavily involved in promoting an emotional disposition, coupled with imaginative and cognitive activity, which has the potential to be converted into tourist activity. Indeed, the activity of tourism itself makes sense only as an imaginative process which involves a certain comprehension of the world and enthuses a distinctive emotional engagement with it. This is true even if the experience of tourism is only confined to a cycle of anticipation, activity and retrospection.

Equally, holiday images can feed back into the imaginative activity of the media. This is suggested here, not just in obvious examples such as holiday programmes (David Dunn), but also in advertisements for Stella Artois beer

(Phil Powrie), the coverage of the Sydney Mardi Gras (Gary Best), the representation of the Third World in the cinema (Felix Thompson) and photographs of holiday activities (David Crouch/Richard Grassick; Claudia Bell/John Lyall). The strength of imaginative media activity associated with tourism may be measured by the growing interest in the notion of post-tourism. As Neil Campbell argues, the mobility of vision offered by the media encourages a new kind of virtual travel along a multitude of paths open to those engaging with representations of the USA.

If we recognize the power of this imaginative force which links tourism to the media, what exactly do we mean by the tourist imagination? We are arguing that the forms and experiences of tourism and the media are substantially distinct even if, as Solange Davin suggests in this collection, they frequently intersect or operate in parallel. We propose the tourist imagination as a bridging concept to explore both the parallels and the differences. It is a notion modelled on accounts of melodrama as an aesthetic mode, conceptualized by Peter Brooks (1995) as the melodramatic imagination, to explore its pervasiveness across theatre, cinema and television in the nineteenth and twentieth centuries. The melodramatic imagination is a model of an imaginative activity that crosses over aesthetic boundaries, particularly useful, therefore, to consider in parallel to exploration of the imaginative dimensions of tourist activities, which pass between different spheres of life. However, it must be recognized that the concept of the melodramatic imagination devised by Peter Brooks is presented as a kind of ideal type which, in practice, may only be partially realized. When we return to a contrast between the tourist imagination and the melodramatic imagination, partial or imperfect realization, or indeed absence of realization, of the ideal type can be just as significant for the interpretation of intersections between tourism and the media. In introducing the tourist imagination then we do so for heuristic purposes rather than as an absolute defining essence of tourism or the discourses of tourism in the media.

The tourist imagination

The tourist imagination as a concept is capable of capturing the mobility of relationships between tourism and the media. It designates the imaginative investment involved in the crossing of certain virtual boundaries within the media or actual boundaries within the physical process of tourism. These boundaries provide many of the familiar dichotomies associated with tourism: work or domestic routine versus travel and holidays; physical restriction or immobility versus movement or virtual movement; a sense of freedom in bodily and mental pleasures associated with travel as opposed to a strict rationing of pleasures required by the needs of everyday subsistence. The importance of the tourist imagination is that of suggesting a creative potential inherent in free movement between different spheres of life. One way of conceiving this creative potential, which has wide currency, is to stress the utopian dimension of tourism or holidays. Thus, at the beginning of his *The Delicious History of The Holiday*, Fred Inglis talks of the cognitive and emotional dimensions of tourism in which holidays are a dream-like emancipation from the world of work. If holidays prefigure utopia then the media play a large part in that kind of anticipation:

> Television is the source of the imagery with which we do our imagining of the future, and the holiday imagery now so omnipresent on the screen – in the soaps as well as the ads and in the travel programmes of all sorts – is one of the best places to find our fantasies of the free and fulfilled life.
>
> (Inglis 2000: 5)

Comparing the role of the tourist imagination in the media with that of tourist activity itself it is clear that the roles are not the same. While both may share the gaze and sound, the direct experience of physical mobility in the activity of creative exploration (with associated smells) is only available through tourism. Tourism involves the actual performance of roles while the tourist imagination within the public discourses of the media can only suggest the possibility of a multitude of roles. Alternatively, with greater scope for fictionalization, the tourist imagination may seem less bounded within the media. Yet media forms, just as much as the actual practices of tourism, are circumscribed in terms of utopian unboundedness, particularly through their own discursive structures and a preoccupation with discourses of conflict. Nevertheless, an appeal to unboundedness must always be there in whatever manifestation of the tourist imagination. This has a multi-directional quality. The imagination is taken from the everyday world into the tourist activity but equally may be brought back from the tourist world as an enhanced imaginative facility. It may be used to appropriate fictions in support of the physical mobility of travel but, equally, travel may be the inspiration of fictions.

In practice it is important to insist that the tourist imagination can never be totally free flowing. While there are always appeals to this ideal of a free flowing imagination, right at the centre of the concept are reasons why it cannot be so. Unboundedness gains its meaning from a promise, but nothing more than a promise, that boundedness can be transcended. We have already noted the role of tourism in offering some kind of movement across major dichotomies of social existence and against the restrictive spheres of everyday life. Yet these boundaries are still there. There is always a recurrent tendency to encounter restrictions which impair the true liberating potential of a tourist imagination. Indeed the juxtaposition of a tourist imagination with non-utopian and conflict-oriented discourses within the media should constantly alert us to the pole of boundedness which threatens to rein in the utopian aspiration. It is in debates about restrictions and unboundedness that much of our understanding of the tourist imagination can be developed.

The tourist imagination in debates about tourism and the media

To examine tourism in this way and the scope for the tourist imagination within the media is precisely to ask about the scope for the utopian in relation to everyday life. As such, the tourist imagination should be seen as a mode of both understanding and feeling about the world which recognizes the utopian aspiration at the individual level but also recognizes that, by the very nature of utopianism, there is inevitable limitation. Thus, the various contributions here can be seen as leaning in different directions – either towards utopian individualism (Neil Campbell, David Crouch/Richard Grassick, Solange Davin) or towards the expectation of utopianism which has not or cannot be realized (Gary Best, Marcella Daye, Rhona Jackson, Felix Thompson). Yet this is not some absolute contradiction. Rather, the appeals by Neil Campbell or David Crouch and Richard Grassick to the axis of freedom in the tourist imagination are simply stressing one term out of the pair which is necessary for understanding of the tourist imagination. The restrictiveness signalled by Marcella Daye or Gary Best emphasizes that the unboundedness, which preoccupies Campbell and Crouch/Grassick, is often denied.

Institutional power in tourism and the media

Contrasting attitudes between unboundedness and restrictiveness then raise a whole area of debate about tourist and media institutions. This includes questions about appropriate methodologies to understand their relationship. For instance, a broadly empirical approach is taken by Marcella Daye. She details the limited range of images evoked by travel writers in their accounts of the Caribbean holiday destinations. Similarly, Gary Best's description of the domestication process by which Australian television attempts to reassure a perceived conservative audience that the gay parade of the Sydney Mardi Gras is legitimate entertainment with the status of a tourist attraction, points up the more radical impulses behind the carnival which are being thereby set aside. Both chapters interrogate the institutional power of the media to produce restrictive definitions of tourist activity. What we have here is a paradigmatic opposition between the power of the media institution on the one hand and the actual tourist sites and practices on the other. This paradigm directly parallels the classic media studies opposition between media institutions and audiences and, indeed, there is a direct overlap, as the readers of the newspapers described by Daye and the viewers of Australian television coverage of the Sydney Mardi Gras are also potential tourists.

The power of tourist and media institutions is also the target of a number of studies in this collection which lean more towards ethnographic approaches in considering how tourists negotiate their own meaning and space. These vary in the degree to which they allow scope for the self-definition of the

tourist. In Sarah Cohen's examination of Beatles' tourism in Liverpool, negotiation occurs within the framework of overlapping economic and cultural
institutions. For Nick Couldry, the rituals of visits to the set of Coronation
Street at the headquarters of Granada Television and the meanings which
individual tourists find there are discussed in terms of the ultimate division of
power between the dominant media institution and the mass audience.
Crouch and Grassick's photography project concerning tourism in the North
East of England is more optimistic, intended to counter the 'dominant media
images of the Northumberland and Durham Tourist Board'. They emphasize
the greater freedom for negotiation against the emphasis on the power of the
media or tourist institutions to restrict the implications of the tourist imagination by imposing categories. As Davin points out, there are parallels in the
disciplines of media and tourism studies in the challenge to passive accounts
of tourists or the media audience who will become tourists.

The industries and their consumers

Davin's argument also raises the question about whether the active audience
or tourist develops utopian aspirations within a reality jointly constructed
by the media and tourist institutions. The parallel discourses and debates
between the disciplines of tourism and media studies are brought together
particularly by the emergence of the hyper-real. Both the tourist experience
and fictional representation in drama heighten our expectations of what
reality should be, causing reality to be reorganized to fit our expectations.
The experience of the media in everyday life tends to converge with tourism
spurred by an overlap between an entertainment industry which creates
images of the tourist destinations as well as the theme park destinations
themselves (Disney). The worlds of tourism and the television thus overlap
as 'a complex web of texts and hypertexts' and it is not surprising that she
points to the possibility that the boundary between television watching and
tourism might almost disappear.

There is then the obvious need further to investigate the connections
between the two industries. But in terms of our investigation of the tourist
imagination, how can the position of the consumers these industries seek
be understood in terms of the two poles of unboundedness and restriction
which we have associated with the tourist imagination? The utopian aspiration which is suggested by the vision of a world re-formed in the image of
the media or tourist promotion, as suggested by Davin, inevitably collides
with the implications of the other pole, the restriction of such aspirations.
In particular, it is interesting to ask what happens when the utopian dimension completely fails – a situation of acute consumer dissatisfaction. Rhona
Jackson examines the role of unpleasure, of a failure of the tourist mode, in
a visit to Los Angeles. In her consideration of a range of analytical
perspectives drawn from media studies on this moment of touristic disenchantment she concludes that the individual failure of the tourist mode

cannot be satisfactorily explained by the more politicised perspectives such as Althusserian Marxism or mass culture theory. More progress can be made by using theoretical approaches which can incorporate the notion of self, drawing on the work of Christian Metz and the notion of the cinematic gaze. Yet in her discussion of the relation between the tourist gaze, as developed by John Urry, in relation to the cinematic gaze it becomes apparent that there cannot be a straightforward reconciliation between central concepts developed in the two different disciplines. There are many aspects of critical discourses used in media studies which have been developed in relation to different objects and therefore cannot be unproblematically transferred to discussions of tourism. The media studies approaches which she applies all insist upon an inherent role for conflict in understanding the media, including the conflicts within the psyche of Freudian theory examined by Metz. Such conflict-based approaches will not be easily assimilated to the pole of unboundedness which we have associated with the tourist imagination.

Discursive structures and semiotic potential

It is also important to recognize that both media and tourist forms have quite divergent implications, arising simply from their contexts within their specific industries and different functions in relation to audiences or tourists. For instance, even in the television holiday programmes from the 1990s discussed by David Dunn, he detects an increasing role for the celebrity system. Performance of media celebrity by presenters, who are supposed to sample holiday destinations on behalf of the viewer, becomes more important than the role of the programmes as consumer guides to holiday choice. This slippage of meaning within the practices of the media will be ascribed to 'semiotic potential', adopting a term introduced by Robert Fish. Such a slippage is possible because, once tourist images are placed within the broader contexts of media forms, there is always more available meaning than can be summed up as purely touristic intent. For Fish, in his viewer-oriented account, the semiotic potential of television dramas in a rural setting is only the start of a process of negotiation. Viewers have alternative choices beyond the stated intentions of the programme makers. They can choose between accepting an idyllic notion of the rural world, open to appropriation by tourist discourse, or a more conflict-based approach arising from the problems of this world which serve to provide dramatic tension. The semiotic potential of rural images can either feed tourist discourse or become bounded within the structures necessary for television drama. Discursive structures of the media can thus provide their own formal constraints to the tourist imagination, an issue to which we will return when considering melodrama in the media.

Consideration of such discursive structures also highlights the methodologies of textual analysis. Even in the more empirical contributions dealing with the role of institutions, close textual analysis, revealing the work of underlying discursive structures, plays a vital role. For Marcella Daye, the effect of Mary Louise Pratt's 'Imperial I/eye' in restricting the range of tourist representations is central to travel writing about the Caribbean. By contrast, analysis of discourses of tourism in the media also shows how tourist images can be easily absorbed in the imaginative processes of the media. Indeed, the tourist associations may shrink to the size of a distant vanishing point. Phil Powrie's textual analysis of advertisements for Stella Artois beer illustrates the way that they filter out the direct associations with the authentic roots of tourist destinations, such as Provence, and the authenticating discourses such as novels or films located in the tourist destination. In the post-tourism of the advertisements what remains is a kind of free scope to draw on these starting points of tourist destination, novels and films as no more than imaginative resources.

But, equally, if the media have the ability to re-accent tourist images according to their own purposes, the very phenomenology of the tourist site can redefine the implications of media images. Tim Edensor's argument about the Wallace memorial considers how the globalized entertainment industry of Hollywood impacts on the experience of the tourist site. The name of Hollywood evokes one whole swathe of possible restrictions upon the tourist imagination although this does not necessarily mean passive acceptance. Just as Neil Campbell points to the possibility of negotiation with media images of the USA in his account of post-tourism, so the visitor to the Wallace memorial negotiates between Hollywood and more local meanings. In both cases, negotiation is within and against bounded options. The tourist has to decide either to accept the globalized media images of Rob Roy or the implications of the layout of the tourist centre with its panorama which suggests further Scottish landscapes to explore. The layout here suggests bodily activity, an engagement with the monument space and the open spaces beyond as an alternative to the mediated melodrama. This is the case despite the presence of videos and other visual material produced specifically for the site which cannot contain the wider matrix of meaning in which the tourist moves. Yet it is also possible that a kind of media production takes place within the tourist site itself which will be part of the multiple flows of meaning from the tourist site and back again.

Media production within the tourist destination has long been possible with the photograph or the home movie camera. Claudia Bell and John Lyall record a new phenomenon of digital production from the mobile phone, enabling the tourist activity to become increasingly intertwined with media production. This allows a third possibility between the potential effects of media categorizations of tourist destinations and the implications of meaning inherent in the geographical positioning and physical layout of the destination. In this case, through the mobile phone, texting, and the mailing of digital images, the very accessibility and immediacy of digital media production allows the tourist to become central to the theatrical drama of their own lives. The flexibility and informality enable escape from the restrictions of collectively produced media forms, even allowing one interviewee to present himself in the middle of active war zones.

Conflict and tourism

Such an escape from the restriction of mass mediated forms only serves to emphasize the boundedness that media discourses operate in general, such that the connection of tourism with contemporary war would be more likely to be considered problematic (even if an option taken by some). As Robert Fish argues, media narratives in such dramas as *Heartbeat* and *Peak Practice* require 'figures of suspicion, derision, deviancy and pity' which are at odds with the requirement to promote idyllic tourist images. Yet, the primacy of the media in handling issues of conflict is not an absolute rule. Tourism may provide alternative frameworks which can accommodate better to conflict than can the media. For instance, in his exploration of Belfast as tourist destination, Kevin Donnelly simultaneously considers the way in which cinema and television have represented Belfast during the troubles. It emerges that aspects of the tourist experience do not appear within the political geography of the Belfast assumed for the dramatic purposes of fiction in cinema and television. Given that such significant parts of the city as Queen's University have never been featured in the fictionalized narratives of cinema and television it is evident that tourist versions can exceed the restrictive definitions of the media representations.

As with the contribution from Felix Thompson, which deals with the intersection between Third World militancy and tourist cinema, the very notion of 'troubles tourism' implies that tourist discourses and the politicized discourses of violent conflict may have productive effects rather than simply being antagonistic. He discusses how the excess of freedom granted to those living at First World standards is contrasted in certain films with a situation of economic and social oppression, effectively a complete anathema to the tourist imagination. Yet in the collision of two apparently contradictory dispositions – between the over-consumption and leisure of tourism or the rigidities of unavoidable struggle against social and economic oppression in marginal Third World communities – significant questions are raised about the political and historical relationship of the tourist to the margins of modernity.

The cognitive element in the tourist imagination

This last argument also suggests why there must always be a cognitive dimension to the tourist imagination. Every individual who looks for unboundedness must make some kind of judgement about how they might cross the boundaries of their own sphere of existence. Such a judgement touches on such questions as what is economically viable for that individual;

of the change in social relationships inherent in becoming a tourist and leaving routine behind; of what will be physically pleasurable and what will not; and, indeed, of perceptions of national identity in many cases. For the virtual tourist, as Thompson argues, the comprehension of the different spheres of existence can be equally important. If we imagine that we cross existing boundaries in order to achieve a mobile vision of the world, then we cannot avoid explaining to ourselves what those boundaries are, even if these explanations are incorrect or ill-informed. Not all is cognitive, of course. There must be awareness of the emotions in departing from the routine and then returning to it. There may be anticipation of memories of sheer physical pleasures of the tourist gaze and occupation of tourist spaces.

The tourist imagination and the melodramatic imagination: media forms and tourist forms

In this exploration of the tourist imagination so far it has been necessary to recognize that the utopian aspiration associated with the axis of unbound-edness in the tourist imagination is always at risk of qualification and restriction. Indeed tourism as an activity is shown to be capable of absorbing thoroughly non-utopian imaginations associated with the troubles in Northern Ireland or wars in the Middle East. In this section the same procedure will be followed by considering a contrast between the tourist imagination and melodramatic imagination as two ideal types which come to intersect. However, in practice, consideration of actual examples must lead to a derogation from the absolute nature of this contrast which helps to suggest some of the contradictory and unpredictable ways in which tourism and the media come to interact.

The differences between the concepts of the tourist imagination and the melodramatic imagination are highly instructive about the kind of absolute contrast of values which may arise in debates about tourism and the media. As Peter Brooks' account of the melodramatic imagination makes clear, the response to modernity encapsulated within the melodramatic imagination is based upon modernity's failures. The melodramatic imagination then is one example of a range of pervasive cognitive dispositions or aesthetic modes central to the media, which are not given to utopian aspiration. Preoccupation with conflict and social failure in such media modes appears antithetical to the pleasurable enticements associated with the activity of tourism. Melodrama, as an example of such modes, will be considered as pervasive in media dramatizations in which the role of conflict is central. It must be noted that the role of the dramatic extends well beyond fiction but may also be used, for instance, in current affairs and discourses of celebrity. It is this widespread resort to the dramatic which provides much of the scope for imaginative exchange with the tourist imagination.

By comparing these two outlooks it also becomes apparent why we might not expect that whole areas of media preoccupation with the melodramatic should be so easily assimilated to the tourist imagination despite the shared imaginative and subjective focus. Peter Brooks' account of the melodramatic imagination is of an aesthetic mode which is a response to a desacralized modernity. In the absence of the certainties of religious belief the protagonists of the dramas he describes are confronted with a world which lacks moral certainty and yet can only have a resolution within the human world. The disasters which befall these characters speak of the conflicts inherent in the onset of modernity from the beginning of the nineteenth century. Of course, the starkness of these conflicts as presented in melodramas often seems too extreme, providing a facile and sensationalist opportunity to personalize evil in a way that has given melodrama a bad name. Nevertheless, Brooks' account points to a deeply embedded framework, even where the more excessive elements of melodrama are absent. This is centred upon the presupposition that a Manichaean divide is in operation, deriving from a conflict-filled and morally polarised world. Central characters are thus faced with the dilemmas of 'moral obscurity' arising from the conflicts in the world. These are explored in terms of their emotions because the moral obscurity does not allow for a rational solution to the problems they face. Instead, typically, the dilemmas of the characters are presented through expressive effects of the drama – most often music in the cinema but also lighting, décor, colour and any other device which can bring to the surface the sense of inner conflict of the characters.

Spectacle from the nineteenth-century stage right up to the twenty-first century blockbuster is at least a part of this expressive emphasis, although there is no doubt that it often exceeds the need to express the dilemmas of the characters. Thus, special effects become a way of selling a film as a kind of marvel rather than expressing the conflicts of the characters. Whether or not the melodramatic dilemmas are accompanied by the excess of spectacle, resolution of narrative conflicts is underlined symbolically, typically by the restoration of the family unit. This symbolic resolution is usually sufficient in emotional terms to close off questions about the underlying causes of the conflict which has challenged the protagonists. It also provides an opening for a more positive interpretation of the drama which may become aligned with the utopian aspirations of tourism, for instance melodramas which deal with hoped-for national transcendence of conflict (Burgoyne 1994).

Broadly then, when considered as ideal types, the tourist imagination can only begin where the melodramatic imagination has ceased to operate because conflict seems antithetical to pleasure and will reduce the freedoms associated with the ideal type of the tourist imagination. Yet, as we have seen, it is still possible that tourism may follow where conflict has been dramatized, as in Kevin Donnelly's piece 'Troubles Tourism' (even though we are immediately aware of a kind of oxymoron which needs explanation). But this can be taken further. Melodrama's preoccupation with conflict can give way to the celebration of spectacle as a marvel in its own right. Its more optimistic use of symbolic resolution may be subsumed within utopian aspirations. Thus, despite the antithetical appearance of the ideal types, it is apparent how exchange may take place between tourist and melodramatic modes. There is plenty of scope for leakage between these two dispositions, either through the coincidence of interest in spectacle, or in the availability of melodrama for the representation of national struggles, as in *Braveheart*, which are commemorated in tourist sites and monuments. This suggests, then, that such contrasts between the tourist imagination and the varying cognitive and aesthetic modes within the media are only a useful first step before considering how derogation from the ideal types actually facilitates a complex series of exchanges in practice. Working back from ideal types to actual practice, then, can provide a means to take further our understanding of other unlikely confrontations and exchanges, for instance between tourist discourses and Third World political discourses (Felix Thompson), or investigations of social problems alongside rural idylls (Robert Fish).

Audience, tourism, performance

A strategy of derogation is also available for actual audiences or tourists. The account of melodrama as a set of dilemmas which require a definitive, if symbolic resolution, suggests that media texts operate according to fixed sets of rules or paths which the audience must expect to follow. However, it is not the case that audiences will necessarily follow any prescribed itinerary of meaning, or adopt a predictable pattern of behaviour. Media studies is thus constantly attempting to adjudicate not just between competing logics within a text, such as the intersection of the tourist imagination with the melodramatic imagination. It also has to address the logics of media texts in relation to the negotiated meanings of the diverse audience. There is always an inevitable lacuna between, on the one hand, text meanings which follow the logic of the text, and on the other, audience meanings and behaviour. Advertising and historical periods of sustained propaganda suggest that media influence can be substantial, but it is never determined in advance by how much and in what ways.

Furthermore, with knowledge of the distracted way in which audiences generally use media texts there are deliberate strategies on the side of producers to address the audience on the basis that their moments of attention may be almost random. Any melodrama which involves extensive spectacle, for instance, may be watched more for this element rather than paltry sketches of the characters. Thus, it becomes commonplace for blockbuster films to be promoted by their special effects rather than by interest in the plot details or motivation. The television text especially has been widely characterized as highly segmented, to give brief rewards of sensation or emotion, making it completely open to the zapping of the remote control. At the same time, and by contrast, television has generated its own fan subcults who know every detail of series such as *Friends, Frasier,* or *Star Trek.*

The distance between audience and text is constantly uncertain and difficult to summarize.

It is not necessary here to follow how these debates have developed in media studies, but rather to note that, by contrast, tourism does not presuppose any fixed texts from which tourist meanings and behaviours can diverge. If tourist studies does not have the text/audience dilemma then it is because tourism is a loose-fit multiple experience without the hermetically sealed temporal 'moment' associated with the notion of a text. There is nothing concrete to refer back to from which tourism starts. Both tourism and media text are enmeshed within numerous flows of cultural events, contexts, desires and feeling. They are erratically mutually informative, with loose and porous borders. The consumer, or 'receiver', of mediated messages acts, ignores, rejects, reacts or negotiates the communicated. But always this means turning back to a sense of the text which prompts these actions. By contrast, tourist practices, while surrounded by a culture of communication, are not required to turn back to a source for their imagination. Beyond the public discourses in which the notion of the tourist imagination has been located here, there are a plurality of possible tourist imaginations which can be put into practice. As Miller argues, the active imaginative consumer participates in 'doing work' through the consumption process. The tourist is poetic, imaginative and creative (de Certeau 1984; Birkeland 1999). These imaginations beyond public discourse are not limited to the detached observance of a tourist gaze, but an active and also physical encounter with the local and intimate worlds that are the content of tourism. The tourist, in building dreams and arranging practicalities, in making journeys and in being there, and then in space/time-reflections, is not identifiable in a tourism 'bubble', but in negotiating, perhaps progressing, life.

Such negotiations cannot be automatically or necessarily derived from the absolutes of the tourist imagination. In order for aspirations on a personal scale to be realized, then, the dominant impact of utopianism in public discourse is often set aside in favour of the practicalities of performance. For the individual tourist always to refer back to the utopian or grander claims of the tourist imagination risks stunting the diversity of tourist imaginations. Recognition of this dimension of performance then requires accounts of why the tourist imagination as a product of public discourse may be set aside or ignored in performance. But yet again, as for audience studies, nothing can be decided in advance about what tourists actually do. In each and every case, as in the studies by Sarah Cohen, Nick Couldry, David Crouch/Richard Grassick the activities of tourist imaginations have to be plotted within a wide range of constraints according to a diverse range of possible negotiated itineraries. At this point then, the tourist imagination must concede primacy to individual negotiation. It would nevertheless be relevant to ask what, if any, part it plays.

Bibliography

Birkeland, I. (1999) 'The mytho-poetic in Northern Travel', in D. Crouch (ed.) *Leisure/tourism geographies*, London: Routledge.

Brooks, P. (1976; 2nd edn 1995) *The Melodramatic Imagination*, New Haven: Yale University Press.

Burgoyne, R. (1994) 'National Identity, gender identity, and the "rescue fantasy" in *Born on the Fourth of July*', *Screen*, 35 (3).

De Certeau, M. (1984) *The practice of everyday life*, trans. Steven Rendall. Berkeley: University of California Press.

Inglis, F. (2000) *The Delicious History of the Holiday*, London: Routledge.

Miller, D. ed. (1998) *Material Culture: why some things matter*, London: Routledge

2 Mediating tourism

An analysis of the Caribbean holiday experience in the UK national press

Marcella Daye

Introduction

This chapter examines representations of the Caribbean tourism experience in two national UK newspapers. Textual analysis of articles on Caribbean destinations appearing in the travel sections of *The Independent on Sunday* and *Sunday Times* for 1998 was conducted. The aim was to identify the main features and characteristics used to construct the tourist experience in representations of the region's landscape. It is suggested that press representations of the Caribbean often lacked distinctiveness in their identity and appeal and tended to promote stylised 'ways of seeing' the region's landscapes and responding to it. This chapter concludes that these representations may be undermining the region's ability to promote a range of touristic experiences and to expand market appeal.

Role of the travel writer and travel writing – mediators of the tourism experience

The media is a primary source of destination images. People also form their images from word-of-mouth stories, advertising and promotions and from their own experiences of destinations. Images of destinations have been classified as organic when messages come from 'non-tourist' sources, and induced when they are formed based on tourism advertising, promotions and campaigns (Gunn 1972; Gartner 1993). The organic images that people hold of destinations are deemed to be more influential than induced images. Inasmuch as they emanate from tourism interests usually associated with the industry, induced images are regarded as less credible than organic images.

However, travel writing in the UK national newspapers seems to straddle these two levels and can be presented or perceived in both guises. For the most part, writers can be categorised as producing induced images insomuch as these articles are often juxtaposed with commercial tourism advertisements and promotions and share their concerns and perspectives. But they also strive to maintain notions of credibility and adherence to journalistic practice. Travel writing is generally not considered to be hard news, but falls more readily into the category of 'soft' feature stories (Keeble 1994).

In his study on the perspectives of national prestige UK travel editors/writers on the nature of their work and its overall role, Seaton (1991:11) observed that they generally felt that it was poor editorial policy to speak negatively about destinations or the holiday experience. The view was that negative or so-called 'knocking' pieces were not particularly effective especially if they were about places people hardly knew anything about. Still, an 'occasional hatchet job' was considered to be good for reader credibility. Space constraints on the travel pages also seemed to encourage more positive news stories, as some travel editors noted that it was a 'waste of space to fill it with negatives when it would be better to give readers good ideas for travel, rather than bad ones'. Furthermore, negative pieces required higher standards of accuracy and more time for double-checking.

In the interviews conducted by Seaton, concerns for high credibility, objectivity or news values, appear not to be high among travel writers, who also felt that readers do not expect these qualities from their stories. In this respect, travel writings in the press are not too far away from their close cousin, the guidebook. Yet, Seaton notes that the travel editors were keen to also maintain their gate-keeping function not to allow overly exaggerated or romantically, inane pieces to slip through onto the travel pages.

The aim of this chapter however, is not to critique the genre of travel writing or travel journalism, but rather to highlight the crucial role of travel writing's unique situation in the national press which gives an appearance of credibility and provides the context to influence readers. Travel writing therefore shares some complicity with the tourist industry, not least of all because of the value of the advertising spend from the tourism industry in the media (Archer 1997; Fry 1998; Holland & Huggan 1998).

There are overt indications of this relationship demonstrated in most travel stories that end with specific information on holiday package prices, telephone numbers and addresses of the relevant tourist boards, tour companies or airlines mentioned in the articles. This information can be seen as a type of direct sales or call to action, by encouraging and, at times, inviting readers to book or call, thereby facilitating opportunities for consumer response and feedback. In a more disguised way, this practice also provides closure for the tourism experience described in the article.

Most of the travel articles in the newspapers reviewed are written in the first person, active voice. This 'Imperial I/ eye' (Holland & Huggan, 1998:15) as first coined by Mary Louise Pratt (1992), not only lends credence to the description of the holiday for readers, but assumes their typical needs and motivations in representing the experience (Seaton, 1991:8). The act of mediation takes place as the travel writer, an apparent independent traveller, finds common ground of interest between the potential tourist and the industry. Deacon *et al* (1999:154) point out that the position of pronouns such as 'I', 'you' 'we' can be studied in media texts to identify 'traces and cues they offer for expressing evaluation of the social reality the text relates to'. In this context the almost universal use of the 'I' by travel writers supports their legitimacy to interpret and construct meaning based on the fact of their actually simulating the role of the traveller tourist.

Locating the Caribbean

In many respects the Caribbean may be seen to be a geographical as well as a socio-political construction. A logical and convenient definition is to describe the area as comprising the group of islands washed by the Caribbean Sea. The problem with this definition is that Bermuda and Cancun are also marketed in the industry as the Caribbean, even though they are not spatially located in the region, and it also excludes Guyana on the South American mainland. The Bahamas too, is not geographically Caribbean, but shares the common history and heritage of the region. Cuba, undeniably in the Caribbean, maintains a unique place, due in part to its political history, and so invokes more complex destination images than just another sun, sand and sea destination. In order to maintain consistency of analysis and evaluation, the sample for this study therefore included countries known as the British or Commonwealth Caribbean that excludes the Spanish, Dutch and French islands of the region.

Analysis of tourism articles on the Caribbean

For the purposes of this study, travel articles on the Caribbean were selected from the travel sections of *The Sunday Times* and the *Independent on Sunday*. Articles were chosen which featured any of the destinations in the prescribed Caribbean area throughout 1998. This yielded 15 articles, nine from the Sunday Times and six from the Independent on Sunday. Ten destinations were featured in the total sample. This sample size was deemed sufficient for the exploratory nature of this investigation, aimed at testing the suitability of coding categories and the overall discursive approach of the methodology.

The content of the articles was also analysed with respect to physical landscape characteristics as used by Dilley (1986:60). Frequency count analysis of these manifest landscape attributes was also conducted to determine the predominant features of the constructs and to facilitate an understanding of the images and patterns of representations of the Caribbean. The articles were also analysed to uncover the underlying meanings of the discourse employed in the representations. According to Tonkiss (1998:253), discourse analysis seeks to 'examine how particular attitudes are shaped, reproduced and legitimised through the use of language'.

In order to determine the constructs of the representations, each of the 15 articles in the sample was coded using a typology of tourist experiences devised by the sociologist Erik Cohen (1979:22). Cohen has proposed a phenomenological typology of tourist experiences that is based on the different meanings which the 'natural environment of others has for the individual tourist'. Cohen lists five modes of experiences: recreational, diversionary, experiential, experimental and existential, and argues that it is possible for some these modes to occur simultaneously. The main implication of these typologies is that people may approach the same landscapes from different orientations and therefore experience them differently.

These modes, according to Cohen, are ranked to represent the range of experiences of the tourists from those who seek 'mere' pleasure in exotic destinations to those who, like pilgrims, search for meaning at the centre of another culture. In summary, the *recreational mode* represents the tourist who does not care for the authentic, but is eager to accept the make-believe in order to enjoy it as a re-creative, entertaining or relaxing experience. Tourists in the recreational mode are so concerned by their motivation to escape that they are relatively indifferent to the choice of a particular destination as long as the basis motivation may be fulfilled. Here, advertisers have a special opportunity to appeal to the recreational tourists in directing their choice towards a preferred destination. The *diversionary mode* tourist is similar to the recreational mode in terms of a desire to relax. However, the main thrust of the experience is predicated by a desire for change and recognition of personal dissatisfaction with everyday life.

In the *experiential mode*, this rejection of the home society is even more intense so that the tourist in this mode engages in the cultures of the destination in order to satisfy his or her 'real needs'. With the *experimental mode,* the tourist sees travel to the destination as a time to closely engage and share in the culture of the host, but in fact makes no real commitment or kinship with them. In contrast, the *existential mode* tourist embraces the world of the Other as real and fulfilling to the point that there may a willingness to live and become a part of the host community. Cohen (1979:17) concluded that it was possible for individuals to engage in various touristic modes as such persons maybe equally at home in two or more 'worlds'.

Discussion of data findings

Dominica and Jamaica were represented in two articles in both *The Sunday Times* and the *Independent on Sunday*. There were two articles on St. Lucia in the *Independent on Sunday* and one in *The Sunday Times*. Barbados was featured twice only in the *Independent on Sunday*. Both newspapers tended to record similar modes of experiences. As Table 2.1 below shows, the recreational and diversionary modes were the most dominant and were often represented together.

The recreational mode is commonly associated with mass tourism which accounts for the majority of packaged holidays (Ryan, 1997). This mode of experience is often criticised for its superficiality and lack of authenticity (Cohen, 1979). Implicit in the recreational mode is the general acceptance that there is a world of make-believe as long as the requirements of entertainment, relaxation or sun, sand sea and sex are realised in the recreational

Table 2.1 Modes of tourist experiences represented in articles

Country	Recreational Mode	Diversionary Mode	Experiential Mode	Experimental Mode	Existential Mode
Anguilla	1	1			
Bahamas	1				
Barbados	2				
Bermuda	1	1			
Dominica	1	1			
Grand Cayman	1	1	1		
Jamaica	2	2			
St Kitts	1	1	1		
St Lucia	2	1	2	1	1
Turks & Caicos	1			1	
Total	13	8	4	2	1

Sample – N=15	
Recreational and Diversionary	7
Recreational, Diversionary and Experiential	3
Recreational and Experimental	2

experience. Cohen argues that the destination or place is not paramount for the recreational mode of tourist experience, as long as this experience is offered. The destination brand image is therefore expected to conform to this mode of experience in order to maintain market appeal (Young, 1999:402). Whatever may be the distinctive features or characteristics of the destination product, it is remoulded and constructed to maintain the consistency of the experience. Li (2000:877) states that geographic consciousness is essential to the tourism experience and asserts that there is a 'spatial-temporal' bond between tourists and destinations.

It was also appropriate to examine landscape attributes presented in the articles in order to determine the most prevalent characteristics. It was found as expected, that landscape imagery was dominated by coastal depictions, especially that of sun, sand and sea. But there were also considerable focus on mountain and scenery, the flora and fauna and rural vistas of the islands.

St Lucia holds the distinction of having all five landscape attributes, with Jamaica and Dominica coming next with four each. For the smaller islands of Turks & Caicos and Anguilla, only their coastal features were mentioned in the articles.

Table 2.2 Landscape representations in articles

Country	Coastal	Mountain	Rural	Urban	Flora and Fauna
Anguilla	1				
Bahamas	1		1		1
Barbados	2				
Bermuda	1			1	
Dominica	1	2	1		1
Grand Cayman	1				1
Jamaica	2	2		2	2
St Kitts			1		1
St Lucia	4	3	3	1	2
Turks & Caicos	1				
Total	14	7	6	4	8

N= 15

Sun, sand, sea – Totally Caribbean

It is apparent that although coastal landscape descriptions predominate, focus is also being directed to other attributes of the destinations such as mountains, rural vistas and flora and fauna. This may be a reflection of an attempt to widen the scope of attractions and vacation experiences offered in the Caribbean. Non-coastal stories included the mountain vistas of St. Lucia, Jamaica and Dominica, bush trails in swamps and mangroves in both the Bahamas and Jamaica and a plantation retreat in St. Kitts.

However it is particularly interesting to note that constructions of the non-coastal landscapes seem to be depicted in much the same language of sun, sand and sea of the recreational mode or experience. In a story of the colonial Caribbean featuring a rest and relaxation holiday on a nostalgic sugar plantation in St. Kitts, the travel writer declared that:

> the entire island seemed to ripple in the breeze…the garden and grounds around the Rawlins plantation house seem more like a 25-acre island hemmed in by a green tide.
>
> (*The Sunday Times*, 29 November, 1998)

Here the travel writer invokes beach imagery in presenting the recreational experience, so that, through his eyes, sugar cane ripples and becomes a green tide. Although the island's black sand coastline was not recommended for swimming, all was not lost for this recreational and diversionary experience, as at the accommodation at Chimney Cottage there was a small, fresh, spring-fed swimming pool, sharing the dining room's view across the sea.

It is clear that this writer actively seeks signifiers that construct the stereotypical, coastal images of the Caribbean. So the landscape is remoulded to suit and to match the main motivations for the travel experience. In the quest to present 'good holiday experiences' that will ensure satisfaction, active construction and remoulding of landscapes are undertaken to suit stereotypical mass market package expectations.

Diving and underwater seascapes were highlighted in the stories on the Turks & Caicos and Grand Cayman. There was practically no mention of land in these articles as for these two writers the sea underworld was the real location for the holiday experience. For the Turks & Caicos Islands, the terrestrial image is of a 'scrawny, sandy place with scrub and fine beaches, which is not a place for the inquisitive'. However underwater it is a completely different scene with 'colourful fish and intense deep blue', which the writer asserts has a 'kind of emotional power, psychological effect all on its own'.

The writer on the Cayman Island continues this imagery by noting that underwater he was surrounded by nothing but 'blueness'. Significantly, these descriptions of underwater seemed to provide opportunities to explore the experiential mode which focuses on a search for interesting vistas and stories. However, the writer describes his introduction to the underwater scooter as 'definitely a touch of the 007 experience in whizzing across the reef' as the highpoint of the occasion, and so maintains the world of make-believe, fantasy and fun which characterises the recreational mode.

Paradise Islands

The stereotypical theme of Paradise and discovery of pristine landscapes is recurrent, especially for the lesser-developed islands. For the island of Dominica, the so-called discovery of the island by Columbus was the lead for the stories in both the *Independent on Sunday* of 20 September and *The Sunday Times* on 22 November 1998. Both writers suggested that should Columbus return today, he would still recognise the island that has managed to escape the ravages of development. Both articles replicate each other in their presentations of an untamed, natural landscape. In the *Independent on Sunday*, the writer mentions 'a tiny fishing village' which has 'mountain frogs the size of chickens and sugar birds, bananasquits which sought out each others reassuring presence'.

There is a similar construction in the article in *The Sunday Times* of 22 November, where, after seeing the 'trees which grow by the roadside like escapees from Jurassic Park and strangler figs that are busy enveloping anything they can take hold of', the writer declared that Roseau, the capital of the island, was a matchbox. Paradise has therefore been constructed by reducing villages and towns to small proportions while conflating the flora and fauna and ascribing human personality to the landscape. This approach in representing landscapes of Paradise by relegating human settlements creates contradictions on issues of development that usually seem to come at the expense of 'naturalness' (Britton, 1979: 326).

Even in Paradise, there are hints of trouble on the horizon. These appear when the experience seems not to conform to the prescribed formula imposed by the writer. From the perspective of both writers, the ubiquity of the cruise ship is threatening the 'naturalness' of Dominica. While the strangeness of nature is benevolent, the appearance of cruise ships in the tiny capital was menacing, so that 'alongside Roseau, the cruise ship 'Fascination' looked as outrageous as a flying-saucer'. The image of paradise associated with a nature island conflicts with overt signifiers of mass tourism, and so the discourse takes on the guise of protecting and

Table 2.3 Terms used to describe landscapes in travel articles

Type of Landscape	Positive	Negative
Coastal – (beach, sea, underwater, diving)	beautiful blue, calm, relaxing, fascinating, cloudless sky, new, deep blue, colour, white sand beaches, clear blue water, plush, palm-fringed, golden sand, lustrous C'bean coast, brilliant bays, unpolluted, spectacular sea	tide of rubbish on beach,
Mountain	lushness, verdant, magnificent, spectacular	too humid, too hot, scrub,
Natural (flora & fauna, rivers, waterfalls, bush, vistas)	barren beauty, paradise, spectacular views, luscious skyline, mesmerising, panoramic, splendid, wholesome, green tide, fertile, ragged, tropical moon, deserted coves, spectacular bushes, rich, spectacular waterfalls, best preserved	horribly overdeveloped, too humid, creepy crawlies, scruffy towns, scrawny prawn, tame, unspectacular, claustrophobic, wet, damp, uneasy place,
Cityscape (urban, modern)	bustling towns, crowded city-centre	Americanisation – shopping malls,

shielding the environment from being inundated with masses of cruise tourists staged here exploiters. Yet this advocacy for the preservation of naturalness has its constraints, when this naturalness with all its wildness and strangeness, is not as friendly and not hospitable enough for the recreational experience to occur. In the same article in *The Sunday Times*, the writer shows some antipathy to the exotic naturalness when it appears out of control. So the volcanic landscape at the heart of the Dominican rainforest, is described as 'hissing and seething' so much that the sulphur steam 'tainted our sandwiches' and made this setting an 'uneasy place for a picnic'.

Trouble in paradise is also the overriding theme of one article on St. Lucia in the *Independent on Sunday* that could be described as a 'knocking piece' and the only dominantly negative story in the sample. Here the recreational and diversionary modes of experience were invoked in the sense that there was a failure to realise or achieve them on the holiday.

Consider, for example, the tide of non bio-degradable rubbish which threatens to engulf many of the beaches and most of the roadsides.

According to the writer, this rubbish is much a part of the identity of the island as the island's stellar volcanic mountains known as the Pitons. The despoiling of the coastal landscape for this writer has totally eroded the experience and there is nothing more to see beyond that fact. Although, he also to proceeds to catalogue a list of problems from harassment to all-inclusive holidays which would make holidaymakers 'disillusioned', it is clear that rubbish acts as a powerful metaphor by which the writer paints the picture of a lost paradise that is no longer pristine.

Representing diversity in the Caribbean

The concept of the Caribbean as a region of diversity with the potential of offering varying holiday experiences appears to pose some difficulties for the travel writers to construe.

Travel writer Hamish McRae in the *Independent on Sunday* of 12 December 1998 wrote:

> As people familiar with the Caribbean know, the key thing to understand about the West Indies is that every island is different. There are big ones and tiny ones: English-speaking, French-speaking and Spanish-speaking; mountainous ones and flat ones; developed and undeveloped; calm and frenetic.

While it is understandable to remind readers that the Caribbean is not comprised of seamless replicas of islands, there appears to be some paucity of words on the part of this travel writer to describe or represent the differences between the countries of the region. He points to landscape differences, language and stages of development to make distinctions between the islands. That some islands are big and some are flat, is not contestable, but here the travel writer reduces their differences so that the reader is unlikely to gain some idea of how these differences relate to the holiday experience.

The tendency to homogenise the region in the construction of landscape experiences relates to an inability to articulate a discourse of difference. According to Hughes (1998), 'such presentations tend to homogenise particularity and in the process corrupt it'. This is aptly illustrated in the way difference or the uniqueness of an island is constructed by attempting to pitch islands against each other. In *The Sunday Times* of 16 October, it was noted that Bermuda's 'edge over the Caribbean is that is works so much better'. But still, compared to the more ruggedly beautiful islands, it was 'more like an obsessively tended garden' that has 'very neat, ridiculously pretty houses'. Here Bermuda's particularity was subjected to what Hughes describes as the 'thematic demands of a touristic mode of consumption'. In explaining the value of a nature based Caribbean experience with no sun and sand and sea, the writer in the *Independent on Sunday* on 20 September notes that the island of Dominica has a delicate eco-system which makes it 'unlike its popular, rum-soaked neighbours of St Lucia and Antigua'.

This ambiguity between 'sameness and difference' as Hughes puts it in representations of the landscapes and space in tourism is also echoed in a critique of the postcolonial discourse of the Caribbean by Iris Zavala (2000:370–373). She asserts that master discourses on the region fail to capture the 'simultaneity of heterogeneities' in constructs of the region. She argues that the overall result of this is the failure of language that resorts to creating a 'fantasy structure, the scenario whereby the society is perceived as homogenous entity'. In relation to representations of the region's touristic experience by mediators such as travel writers, differences and particularity are not significant or considered experiences to be explored in detail. Their emphasis, as shown from the analysis, is on maintaining stereotypical constructions.

Deconstructing Caribbean landscape representations

In accounting for the tendency in stereotypical representations, a valid explanation is that of the commodification of places in tourism. The concept of the tourist gaze as expounded by Urry (1990) maintains that the appropriation of places by the industry reflects the consumptive behaviour of capitalist economies. This perspective centres the argument on landscape representations in terms of attributes and their functional use. Therefore, the coastal landscapes of the Caribbean are evaluated in terms of their benefits in providing recreation and even hedonistic pleasure to tourists. But there is a risk of limiting the meaning of landscape representations only in terms of functional explanation of attributes.

For example, Echtner & Ritchie (1993) point out that besides an appreciation of the attributes of a destination, people also have holistic images that relate to their psychological, emotional and attitudinal responses. It could be further argued that landscape representations are endued with cultural and social meanings that are not simply for consumptive appropriation, but also set the terms and parameters for their own evaluation. As Bender (1993:1) writes:

> the landscape is never inert, people engage with it, re-work it, appropriate and contest it.

The process of representing the scenic beauty of tourist destinations involves an active construction of the landscape in terms of aesthetics and values that reflect Western ideals. In his work, *The Aesthetics of Landscape,* Bourassa argues that landscape tastes and preferences reflect the attitudes and biases that people bring to them. The landscape is 'one form through which cultural groups seek to create and preserve their identities' (Bourassa, 1991: 91–2). He identifies familiarity as a key source of cultural aesthetic attitudes that gives meaning to experience of places. Therefore, landscape stereotypes can be seen as part of cultural strategies that reinforce the familiarity required by tourists to negotiate and manage the landscape of the Other (Dann, 1992). The landscape representations of the Caribbean in travel writings in the national newspapers serves this role of familiarity, and is successful to the extent that there is the power to create, ascribe and script tourist experiences of the landscape for their readers.

Conclusion

This chapter has examined representations of the Caribbean holiday experience in the UK. The findings show that the recreational mode of tourist experience is dominant in landscape representations. It has been argued that although there is an overall emphasis on the coastal attributes of the countries of the region, increasingly, other physical attributes such as mountain vistas and flora and fauna are being featured. In spite of this development, the representations of the region's landscape are constructed to depict stereotypical, recreational experiences. The way of presenting remains the same, even with shifting vistas and landscapes. The writings of the travel press on the Caribbean therefore tend to reduce the particularity and diversity of the region's landscape by emphasising and legitimising a preferred way of seeing and experiencing the vacation experience.

I am proposing that media representations of the holiday experience in the Caribbean reflect a closed cognitive orientation that tends to reduce or to reconstruct various ways of presenting or experiencing the region into prescribed stereotypes (Young, 1999: 402). These constructions appear to be an expression of individualistic and shared meanings that prescribe values to particular experiences (Squire, 1994). I am not suggesting however, that these representations by themselves account for, or determine the choices of destinations by tourists or the subsequent behaviour on holiday. However, the argument here is that these media texts appear to be mostly prescriptive in their construction of these representations and experiences. In this regard, the media plays a significant role in moulding the motivations, perceptions and the demand of tourists, and, in turn, the nature of tourism development in the Caribbean.

Bibliography

Archer, B. (1997) 'Sun, Sea, Sex, Sand & Media', *Campaign*, 29–30.

Bender, B. (1993) *Landscape – Politics and Perspectives*, Oxford: Berg Publishers.

Bourassa, S.C. (1991) *The Aesthetics of Landscape*, London and New York: Belhaven Press.

Britton, R.A. (1979) 'The Image of the Third World in Tourism Marketing', *Annals of Tourism Research*, 6: 318–329.

Cohen, E. (1979) *A Phenomenology of Tourist Experiences*, Aix-En-Provence: Cedex Centre des Hautes Etudes Touristiques.

Dann, G. (1992) 'Travelogs and the Management of Unfamiliarity', *Journal of Travel Research,* 30: 59–63.

Deacon, D., Pickering, M., Golding, P., & Murdock, G. (1999), *Researching Communications*, New York: Oxford University Press.

Dilley, R. (1986) 'Tourist brochures and tourist images', *Canadian Geographer,* 30: 59–65.

Echtner, C., & Ritchie, J.R. (1993) 'The Measurement of Destination Image: an Empirical Assessment', *Journal of Travel Research,* 32: 3–13.

Fry, A. (1998), Fantastic Journeys, *Marketing,* 26: 35–36.

Gartner, W.C. (1993) 'Image Formation Process', In M. Uysal & D.R. Fesenmaier (Eds.), *Communication and Channel Systems in Tourism Marketing*, London and New York: The Haworth Press.

Gunn, C. (1972) *Vacationscape. Designing Tourists Regions*, Washington DC: Taylor and Francis/University of Texas.

Holland, P., & Huggan, G. (1998), *Tourists with Typewriters – Critical Reflections on Contemporary Travel Writing*, Ann Arbor: The University of Michigan Press.

Hughes, G. (1998) 'Tourism and the Semiological Realisation of Space', In G. Ringer (Ed.) *Destinations: Cultural Landscapes of Tourism*, New York: Routledge.

Keeble, R. (1994) *The Newspapers Handbook*, London and New York: Routledge.

Li, Y. (2000), Geographical consciousness and tourism experience, *Annals of Tourism Research*, 27: 863–883.

Pratt, M.L. (1992) *Imperial Eyes – Travel Writing and Transculturation*, New York and Canada: Routledge.

Ryan, C. (1997) *The Tourist Experience: A New Introduction*, London: Cassell.

Seaton, A. (1991) *The Occupational Influences and Ideologies of Travel Writers, Freebies? Puffs? Vade Mecums ? or Belles Lettres ?* London: Centre for Travel and Tourism with Business Education Publishers Ltd.

Squire, S. (1994) Accounting for cultural meanings: the interface between geography and tourism studies re-examined, *Progress in Human Geography*, 18: 1–16.

Tonkiss, F. (1998) Analysing discourse, In C. Seale (Ed.), *Researching Society and Culture*, London: Sage Publications.

Urry, J. (1990) *The Tourist Gaze*, London: Sage.

Young, M. (1999) 'The relationship between tourist motivations and the interpretation of place meanings', *Tourism Geographies*, 1: 387–405.

Zavala, I.M. (2000) 'The Retroaction of the Postcolonial-The Answer of the Real and the Caribbean as Thing: An Essay on Critical Fiction', *International Journal of Post Colonial Studies*, 2: 364–378.

3   Media Makes Mardi Gras Tourism Mecca

Gary Best

Image of desire

I walk by the STA travel agency in the University's central plaza and see the image of a shirtless man with 'Nice package!' in white letters across his bicep and under his pectorals. Smaller print informs of 'Mardi Gras Packages to accommodate all tastes'. Great looking guy! Great strategy!

The even finer print across his white shorts inventories obvious Mardi Gras content – in upper case: DISCO SEQUINS G-STRINGS PRISCILLA DARLING DRAG QUEENS PARADE DISCO DANCE, and lower case: dance party floats boys boys boys glitter glamour pride marching boys frocks parade disco sequins g-strings Priscilla fabulous oxford street dykes on bikes. Unpunctuated, the stream of consciousness style mixes rhetoric and reality and, apparently like Mardi Gras, flings aside conventions. The predictable inventory includes many stereotypical descriptors pertaining to hetero-hegemonic perceptions of Mardi Gras and gay men partying. The under/non-representation immediately apparent is the 'lesbian' component of Mardi Gras; one reference to dykes on bikes hints that the Sydney Gay and Lesbian Mardi Gras (SGLMG) is very gay, but lesbian-lite. The single lesbian reference may not only allude to a serious counterpoint but also a problematic gay men/lesbian polarity, as well as the issue of in/visibility. The STA poster is all about another construction, that media darling, the highly visible Mardi Gras *man*.

Once vilified, the SGLMG is now celebrated in dedicated newspaper supplements, TV telecasts and tourism publicity, and publicised as an international event of economic and social significance. The media, and mediation, have contributed to the creation of all that now is Mardi Gras. But how has such a phenomenon, and such an obvious, dramatic shift in attitude, both public and media alike, come about?

Differentiation and dualisms

Any encounter with the SGLMG parade provides a heightened sense of a complex realism. Part of such complexity is the ways and means that Mardi

Gras differentiates itself from its operational context, and that context's range of public response via the media. This differentiation is constructed on a duality that privileges dialogues of separation and distinction:

> A focus of gendered/sexed and sexualized embodiment foregrounds subjectivity as always fractured and multiple, and contests hierarchical dualisms such as mind/body, Self/Other, gender/sex, tourist/host and straight/gay.
>
> (Johnston 2001: 181)

The Self/Other and straight/gay have been the primary underpinning of the SGLMG since its initial protest march form in 1978. Twenty years later the parade was, according to then president Julie Regan, essentially a civil rights demonstration (Fyfe 2001), yet such a perspective continues to be largely ignored as the media persists with its focus on both the theme, and images, of gay men partying. The Self/Other and straight/gay dualisms are still profoundly evident and, rather than contribute to any articulated attempt to dismantle and/or deconstruct such dualisms, the media continues to nurture the Other as a primarily gay construct.

> The SGLMG's Otherness has been evident since that initial protest when, as a phenomenon, it first claimed space that has been tightly held annually in a number of ways. There is, however, no single occupying entity, despite the convenience of the all-inclusive operational title.
>
> (Ryan and Hall 2001: 115)

The space is contested, and because it is also the annual site of an event that privileges a particular form of display that is usually invisible because of its Otherness, its drawing power becomes magnified. It is an opportunity to be somewhere extraordinary, even if only for a few hours each year. Space ordinarily accorded 'public' status is appropriated on the Parade day, but it is an 'authorised transgression' (Seebohm 1994, citing Eco, p. 202), in that permission has been granted by authorities that act to regulate and control. In the case of the SGLMG:

> It must be remembered that this 'licence' [is] to make use of public space for the open expressions of (homo)sexuality, thus enabling gay and lesbian participants to wear flamboyantly outrageous (or, in many cases, very little) dress, and to openly show signs of affection and passion.
>
> (Markwell 1998: 117)

The search for the extraordinary so often urges the tourist on. A new space, a dynamic (urban) space such as that characterised by Knopp (1995) with its contingent interconnections, resistance to reduction and spatial dynamism, a new, unmapped territory, lures the tourist/ spectator.

An article focusing on Sydney during the 2000 Olympics was titled 'Fun and Games: In Sydney, the vice is widespread and legal'. Its theme was essentially 'Sydney's licentiousness', which was characterized as 'institutionalised'. As Horowitx (2000) notes, the annual Gay and Lesbian Mardi Gras, a wild, flesh-baring carnival, is partly sponsored by the government and business sponsors such as Qantas ... The Mardi Gras also garners the largest TV audience of any event in Australia.

Here the SGLMG becomes a government supported and condoned Bacchanal, a view not inconsistent with the majority of Australian media representations, including the STA imagery, employed to lure tourists to Mardi Gras. This view is also largely undifferentiated; Mardi Gras is image, destination, event and party, foregrounding the exposed and sexualized flesh, only addressing the politics when absolutely necessary. Such necessity may be required in order to legitimize attendance or reporting of the event; it becomes Other to the Bacchanalian revelry. How does the reporting work? How is the knowledge disseminated? Who watches? Why watch? Why attend?

The two media that provide most coverage of the SGLMG and parade are print and TV, although the Internet now offers the most focused and extensive information provision. Whilst there have been significant changes in both approach and content since 1978, the imagined status quo of the Other appears to be much in place, and annually perpetuated by the media.

First, some fine print, with images of desire and discomfort.

On the Monday following the 2001 parade, Melbourne's *The Age* published a photograph of an elaborately costumed man, described as 'a participant', and made scant reference to the Mardi Gras itself, but noted that 'Hundreds of thousands watched' (*The Age* 2001). Melbourne's print media position is one of distant acknowledgment, an event of the Other in a city which is also Other to Melbourne.

In *The Sunday Age*, an article on two of the Melbourne Marching Boys (SGLMG Parade regulars), concentrated on their ritual waxing, pumping and tanning preparations for the big night (Fyfe 2001). The expectation of the gaze is so fundamental to the Parade that the media emphasis on the physicality of it all has, in recent years, often overwhelmed the themes, the issues, and the now diminished problem of protest. The rhetoric of that protest, and the violence that marked the embryonic 1978 March have all but disappeared, conveniently replaced by much more marketable flesh, feathers and fantasy.

Two versions of a 2001 publicity photograph showed one of the 2001 TV broadcast hosts, Rove McManus, looking uncomfortably at drag queen Magnolia. The photos position the Otherness of the Mardi Gras and its participants when placed beside the non-Mardi Gras world; Mardi Gras reality disconcerts and challenges the familiar, non-threatening reality of Rove. In the accompanying article, Rove articulated a predictable interest and support for the event rather than the evident discomfort displayed (Hood 2001). A more provocative version of the photograph used in the dedicated gay press *B.NEW.S* (1 March 2001, p.17) showed a startled Rove with Magnolia on his left, curving her tongue towards his ear; that lascivious tongue was, clearly, just too much for mainstream publication.

The Sydney Morning Herald (SMH) of 9 February 2001, fully embraced the Other as its own in the twenty-page supplement *The Sydney Gay & Lesbian Mardi Gras Festival Guide*. On a pink background, the headline 'Who's coming out for Mardi Gras' above a bald-headed male (?) gazing through a sequined mask resonates with yet more stereotypical mainstream Otherness. Content-wise, the guide provided reviews, gay-themed ads, and a festival planner revealing that Mardi Gras is much more than a parade, yet it is the parade that continues to get the press.

The Sydney Morning Herald and Mardi Gras have not always enjoyed such mutual admiration. Carbery's history of the SGLMG (1995) reveals the *SMH*'s shift from opposition (twice publishing the names, addresses and occupations of those arrested after the 1978 March) through misre-porting, or non-reporting, in the mid-1980s to acknowledgement in a 1994 editorial that Mardi Gras had a right to determine party entrance policies. Altman (1997) observed that the *SMH* was sufficiently emboldened by the early 1990s to publish a Mardi Gras crossword, but twenty years earlier had refused to review his first book *Homosexual: Oppression and Liberation* (1971). It is tempting to believe that the *SMH*'s very gradual moves towards acknowledgement and eventual support were slower than community acceptance generally, given the more rapid and obvious growth in spectator numbers, but the *SMH* experience probably provides a more accurate measure of the pace of so-called mainstream acceptance, at least in the Sydney context. Three print media responses that succinctly illumi-nate the nature of the relationship in Sydney also have a broader relevance:

> Over the last three decades of the 20th century, the Sydney straight print media has treated the gay and lesbian community with a mixture of hostility, ... benign and titillated amusement, and invisibility.
>
> (Scahill 2001: 189)

Hostility, benign and titillated amusement, and invisibility are possibly the three most accurate descriptors of mainstream response to the SGLMG. The collapse of the Satellite Group (gay media and property) highlighted the context of wider media reportage of gay issues and events, and the almost inevitable schism between dedicated gay community content and the approach of mainstream print media to gay content:

The decreasing bigotry of the general public and increasing integration of the gay and lesbian communities have prompted some questions about the relevance of the gay and lesbian press.

(Powell 2001: Media 14)

If there is a 'decreasing bigotry' and an 'increasing integration' then perhaps spectator crowds at the SGLMG provide a measure of such decreases and increases, although attendance can represent the entire spectrum of response from protest to total support. The SGLMG may be 'almost universally lauded' but a column in the *Sydney Star Observer* (22 February 2001) disengaged from 'this so-called community celebration', and lamented the family flavour of the Mardi Gras parade. Powell also quotes Michael Woodhouse of the AIDS Council of NSW who acknowledges that 'more and deeper stories on homosexual issues than ever before' are run in the mainstream press but that the detail is still provided by gay and lesbian newspapers and magazines. The responsibility for the detail may always remain with the dedicated press; the mainstream will probably only continue to skim the surface, or only be *permitted* to skim the surface.

Virgin Blue's in-flight magazine, *Voyeur*, highlighted the dancing, and the flesh, describing it as 'the largest celebration in Australia, so get along to Sydney this Mardi Gras season and soak up the fun, glitz and glamour, darling' (Clark 2001: 11). The discussion is accompanied by images of dykes on bikes, as well as the Sisters of the Order of Perpetual Indulgence, gold-lamé-clad drag queens, the caption proposing, unproblematically, that 'The Mardi Gras attracts people from all walks of life', the images offering something for all the family flying with Virgin. Accordingly, it seems that:

The Australian community has become very accepting of gay people. Remember when the Mardi Gras was a protest march, the idea was that gay people wanted to be part of the mainstream, and now that's happened. There is still a gay culture, but gay people are also much more part of the community.

(Overington 2001)

Every Mardi Gras parade manifests pleas for equality, but not necessarily inclusion. Parts of state capital cities may provide spaces that have become identified with gay life, but the spaces remain marginalized and unassimilated. The Mardi Gras parade may foreground what appears to be a celebration of identity and progress, but the agenda of protest remains. As usual, and despite the fact that the article concerned the 2002 Gay Games, it was accompanied by an image of two slim young men in loincloths dancing on an unnamed Mardi Gras float. The print media continues to characterise 'the gay experience' with images of scantily clad dancing boys. Needless to say, such an image probably has become, for many, what the SGLMG parade is all about, and certainly links to the STA 'Nice Package'. The image's caption is: 'Mainstream: Sydney's Gay Mardi Gras' which not only echoes one reading of progress consistent with that articulated by Fisher but also locates 'Sydney', constructs 'gay', and excludes through inaccurate quotation 'lesbian'.

Desire to attend the SGLMG (Parade)

It is here that the convenient construct of the gay male tourist begins to emerge as both Self *and* Other. If one 'identity' has developed as a focus for target marketing, the gay male tourist appears to be the likely candidate, but there has to be more to the basis of appeal than to attend a four-hour parade. What is such an attraction's appeal? Whilst Pearce, Morrison and Rutledge (1998) address what they term 'sustainable tourist attractions', the SGLMG relates directly to 'clear symbolism to reveal the attraction's theme' and 'public understanding'. Operational requirements of the former seem to be more than met in the case of the SGLMG, proclamation being one of the parade's obvious strengths. The second observation leads into a brief discussion of Australia's fondness for 'big', much larger than life representations of the ordinary that transcend their ordinariness through unexpected scale. The SGLMG is already the biggest dedicated gay and lesbian parade in the world, so scale is certainly addressed.

Appreciation and understanding of the object of the visitor's gaze will inevitably vary, and with the SGLMG parade, the visitor's level of interpretation will vary according to their level of commitment, experience, motivation and desire to be of the parade, or remain an observer. The SGLMG gaze is multi-directional, a condition true of most parades but the spectacle in this instance is more provocative than most. Images are 'of paramount importance because they transpose representation of an area into the potential tourist's mind and give ... a pre-taste of the destination' (Fakeye and Crompton 1991: 10). The visitor attraction market is explored by Swarbrooke (2002), and two stages of the discussion suggest potential for future research on the SGLMG – the nature and process of individual decision-making, and market segmentation in relation to the Mardi Gras 'product'. Whatever does inform the decision to attend, the figures are impressive:

> There were more than 5,000 international visitors to Sydney during the 1998 festivities, 3,600 who came specifically for the event. Of the 7,300 interstate tourists, 4,800 had Mardi Gras in mind ...
>
> (Ryan and Hall 2001: 111)

For many commentators, observers and participants, one fundamental question forms: is the SGLMG a celebration of sexuality, or an event that is more closely aligned with considerations of sex tourism? (Ryan 2001). For many, 'the wild flesh baring' of the parade might seem to suggest that there is an equally public availability of sex as a direct consequence of the parade.

The post-parade party had, by the mid-1980s, a reputation as 'a Bacchanalian orgy' (Carbery 1995: 73) but its popularity with 'straights' in that decade and into the 1990s limited its appeal for the gay/ lesbian/ transgender community as a post-parade destination.

Sex tourism has been defined as tourism essentially for commercial sex purposes but that appears too simplistic. Subsequently, other variables or parameters have been proposed that further illuminate the complexities of the concept of sex tourism: travel/ length of time/ relationship/ sexual encounter/ and those involved. To understand sex tourism, the travel behaviour, and the parameters that delimit that behaviour, must be located in broad societal trends (Oppermann 1999). One means of distancing sex tourism is to marginalize those who might appear likely to only engage, or involve, moving the margin to the 'fringe' of society. If it is the mainstream that defines the edges, such definition must create challenges and opportunities, such as challenges to such disparate yet parallel concepts of hegemony, monogamy and hedonism. If such a premise is to be accepted, then there is likelihood that sex tourism may be one sub-operational element of the SGLMG. Despite the temptation to do so, however, the overt sexual nature of much of the Parade's content need not necessarily suggest causality in terms of sex tourism activity. The SGLMG parade is public provocation and spectacle but entering or participating in either the lifestyles or activities of that zone of provocation is not a given.

Sydney's high profile gay/lesbian/transgender community has made it a popular destination for those seeking specific travel outcomes, sexual or not. The SGLMG has, over the last decade, become the jewel in the crown of travel packages built around the event. Mardi Gras Travel (www.redoyster.com) continues to offer a range of packages – Gold, Silver and Bronze passes, Cairns, and SPIN-FX Desert Dance Party Package at Ayers Rock. Above and Beyond Tours (www.abovebeyondtours.com) features Sydney's Mardi Gras, and Silke's travel (www.silkes.com.au) claims to be Australia's Premier Gay and Lesbian travel site, offering Mardi Gras packages, gay and lesbian harbour cruises, and wine tasting. STA Travel (www.statravel.com.au) offered 'Mardi Gras packages to accommodate all tastes': The full number, The three-piece, The bare minimum, Recovery at Palm Cove, and SPIN-FX, The Desert Party. The associated images are always resonant – the promise of travel in the time of Mardi Gras is all about the package which, in itself, seems to promise more than the sum of its (exposed) part(s). The travel agency advertisements often blur gender alliance in perhaps a more acceptably hegemonic manner than the Parade imagery – they 'look' like us, they might just be friends yet in the context of Mardi Gras such a reality might be hoped for but is unlikely. Spaces such as beaches are as mediated as the Parade route in that regulation allows transgression within a time frame; certain behaviours may be tolerated only during Mardi Gras, especially if the space is primarily coded for public activity. Whilst a territorial imperative may be manifested in terms of numbers, any diminution in those numbers usually means a reversal to usual operational states, the prime example being the parade route of the SGLMG. That parade route has become one of the best known in Australia via various live and taped broadcasts on TV, the other media force.

Desire on the TV

It's difficult to describe those times now, because we've seen the Mardi Gras on TV and it all seems so big and brilliant, but it wasn't always like that.

(Yang 1999)

The SGLMG has been televised since 1994, first by the Australian Broadcasting Commission (ABC) from 1994 to 1996.

The ABC's decision to televise Mardi Gras in 1994 was both controversial and successful, at least in terms of ratings. But the opposition – including a petition from a large number of parliamentarians asking it not to be shown until late at night – showed the ambivalence of middle Australia.

(Altman 1997)

In 1997 the parade was covered by the ABC's McFeast, a current affairs show hosted and written by Lisbeth Gorr as the eponymous Elle McFeast. As a forum for Mardi Gras, McFeast offered both context and containment. Elle McFeast's approach to Mardi Gras was hardly ambivalent but contributed little that was new or unexpected. McFeast's Mardi Gras focus was on the people as well as the event, although each individual was in some way of the event. One council worker coyly indicated that 'it should stay home in the bedroom – in the closet', but most offered messages of support, that Mardi Gras was all right, and doing wonderful things, although what exactly the wonderful things were did not make the final edit.

The parade coverage included many images widely associated with Mardi Gras – bare breasts and behinds, close-ups of the grabbing of ample crotch packages, comedic simulated anal intercourse, and associated bending-over jokes, Flight Attendant Phillipa Hole of Gayviation, the large walking prophylactic that announces 'It's fuckin' hot in a condom', and so on. The images were predictable and reductive, scandalous yet familiar; the commentary only reporting what was already obvious. Of course, the appeal for many in seeing such imagery on the TV should not be overlooked. These images may be, in one sense, what many attend the parade to see, and watch the TV coverage to see – public displays of the Other in, for a specified period, uncontested space. There is a safety in watching in numbers – the crowd as voyeur, but Urry's index in The Tourist Gaze (2002) does not include voyeurism under 'V'. The parade, it seems has the purpose of secular display. It may not, however, be any more accessible to the watching tourists than an extreme religious ritual. The cultural context of 'queerness' surrounds and arguably inhabits the paraders, but to some extent also extends to the bodies of the watching tourists.

(Johnston 2001: 189)

McFeast acknowledged in passing the aspect of the Mardi Gras parade that continues to foreground the gay and lesbian political consciousness of the first protest of 1978. Issues of gay and lesbian parenting and families, equal civil rights, the fight against ongoing discrimination, gender and trans-gender, and those who march in Remembrance of those lost to HIV/AIDS – which is really all parade participants, as well as those in the specific parts of the parade.

Channel 10's coverage of the 2001 SGLMG parade took a very similar form, apart from the compulsory warning at the beginning – M (Mature) rating, and that nudity and some coarse language were likely to be encountered. The presenters were Julie McCrossin, Rove McManus, and drag queen Vanessa Wagner. Rove confided that he had consulted both homosexual and heterosexual friends for special terms to describe the event and the mood – 'fabulous', 'gorgeous', and 'marvelous' led his predicable lexicon. When the commentary turned to discrimination, Rove volunteered: 'As a heterosexual male who runs like a girl, I know what it's like to be discriminated against although not as much as these people – but look at them, though – they're having a ball.' Distancing from the Other, gazing upon the Other but no engagement with the Other – Rove determinedly maintains his heterosexual space.

Julie McCrossin offered observations that indicated either an attempt at more insight, or a better script. Early in the parade she characterised it as 'an enormous indicator of social change' and whilst it certainly is some kind of indicator, its place on any scale would register high in Sydney, relatively high in most state capitals but rapidly diminish away from those few cities. When first broadcast by the ABC the backlash came not just from the opposition in Parliament, but also from rural locations where the only TV is the national provider. On the ABC's *Backchat*, a feedback show that provided voices for written opinion, the responses ranged from outrage to conspiracy theories of the Federal government being populated entirely by homosexuals. McCrossin claimed: 'Here in Australia we're becoming almost used to the idea that once a year you have this amazing display of gay and lesbian life on the streets.'

Her qualifier 'almost' is significant, as it is still difficult to imagine coverage of the SGLMG parade being warmly welcomed into most suburban, or rural, lounge rooms. Knowledge of the parade is one thing but to watch it with the kids is quite another. McCrossin's reference to social change is accurate in the sense that there has been a significant increase in the dissemination of sexual facts in schools and communities as a direct result of HIV/AIDS education. Whether there has been a similar increase in awareness of, not to mention tolerance and acceptance of, the gay/lesbian/transgender community is much more difficult to establish. The edited coverage presents the community, with its diverse lifestyles and open sexuality, as fact – here it is, celebrate with us, look and learn – but in case the meanings are not clear, or should the viewers need something of a context that is familiar and non-threatening, here are Rove and Julie to help out.

The third commentator was drag queen Vanessa Wagner, also an (urban) media presence via her commentary on a number of local TV shows, her presence in advertising for a local Internet Service Provider (where her 'real' male name, Tobin Saunders, is provided as well for the customer uncertain about subscribing to a drag queen's ISP), and her personal approach to drag of not shaving facial, chest or leg hair. Vanessa's interviews provided bites of recognizable, mainstream performers, none of whom threatened with political advocacy, exposed flesh, or simulated sex acts. These touchstones of normality could possibly reassure that all was not lost in the maelstrom of Mardi Gras.

There was much that was political in the coverage. The theme of the 2001 parade was gay and lesbian parenting, so the pink picket fence provided the kick-off for Children of the Rainbow, Rainbow babies, Planned Wanted Loved (PWL), and Parents and Friends of Lesbians and Gays (PFLAG). Political parties and politicians, ethnic, religious, and community groups, and gay/ lesbian/ transgender special interest groups such as Sydney Leather Pride ('the group that every cow fears', according to Rove) and the Order of the Sisters of Perpetual Indulgence, represented by Sister Mary Mary Quite Contrary and Sister Mary go Round (Seebohm 1994: 211–212 on status reversal, and unification of opposites) all paraded to celebrate their presence. Whilst there were some bare breasts and buttocks, some provocative moves, and one float consisting of giant male genitalia, most of the parade's visual content was similar to that of previous years. Given the broadcast history of the parade, much of what is now encountered on an annual basis by a viewer who has been there since the initial broadcast on TV must look very familiar. There is, of course, on-going debate within the gay/ lesbian/ transgender community about the nature of what is both included and presented in the parade. Markwell (1998), like Ryan and Hall (2001) notes that some gay men and lesbians are offended by what they consider to be stereotyped, even incorrect portrayals of their sexuality (p.115).

There are also those who still protest on the same annual basis, notably the Reverend Fred Nile and his Festival of Light supporters, always praying for rain to wash away the parade and the sins of those participating. Naturally there has been an on-going conflict between Reverend Nile and the Sisters of Perpetual Indulgence, most notably in 1989 when the Sisters carried a huge papier-maché head of Fred Nile on a platter of fruit (Carbery 1995: 110; Seebohm 1994: 209).

Religious protest was picked up as a theme by Robert Hughes in his personal documentary series, *Beyond the Fatal Shore* (2000). In the first episode, Hughes counterpoints hedonism with wowserism, the wowser being 'a puritan of advanced prejudices, a killjoy, a 'blue-stocking" (Baker 1966: 136–138). Hughes sums up the SGLMG:

> To look at Australian pleasure a good place to start is Sydney's annual Gay and Lesbian Mardi Gras. It's the extreme public expression of our national hedonism for gays and straights alike. It attracts well over 500,000 and a few years ago it actually outdrew the Pope's visit to Sydney. It would have been unimaginable in the Australia of my youth.

For Hughes, the SGLMG is all about pleasure, and attempts to thwart the pleasure are met with stern disapproval:

Back at the build up to the Mardi Gras, the wowsers are out in force. They're here every year and they're something of a sideshow in their own right.

ROBERT HUGHES:	What do you hope to achieve by coming here?
WOMAN 1:	It's simply to speak out for righteousness; they've been promoting this as a family day.
ROBERT HUGHES:	Do you think that all the parents who come with their children are in some way morally deranged?
WOMAN 1:	Yes, I do.
WOMAN 2:	It is a combination of the laxness of Australians. They'll be – she'll be right, let's have a good time.
ROBERT HUGHES:	Well, I would have thought that was what this was about, it's not about evangelism, it's about having a good time ...
	In a gesture of pure wowser killjoy, the soldiers of Jesus usually pray for rain to dampen down the ardour of the proceedings.
WOMAN 1:	I would love the Lord to rain on this parade, yes, I would. I absolutely make no excuse for that.

Praying for rain is the primary, and apparently only, strategy of those who protest against the SGLMG. Rain would begin to serve their purpose but in hoping for more, the Reverend Fred Nile made the compassionate point that:

> We have no option but to ask God's intervention to stop the Mardi Gras with something like thunderstorms or hailstones – but of course not big hailstones, not large enough to actually hurt people.
>
> (Carbery 1995: 125)

The essence of the SGLMG for most, it appears, is the desire to have fun but the nature of that fun is as diverse as the parade itself. Hughes may be technically correct when he claims the SGLMG is not evangelical, but the parade, its participants and spectators have a gospel of sorts that is distinctively their own.

Destination of desire

The concepts of destination image formation and travel experience resonate in this context. Ross (1994) cites Gunn's seven phase model of travel experience but what the model does not consider is the challenges facing those who seek images and information that may not be easily accessed, or have the perception of sub-cultural stigma associated with them. How does the tourist whose sexuality is unresolved, or not public knowledge, gain the specific knowledge when the very act of accessing may lead to undesirable outcomes of association or exposure? The nature of accessing information can take many forms, some that may threaten the individual's status quo. Gunn's assumption is that the search for both image and information are public and unproblematic but in the case of the sub-cultural, such an assumption is invalid. The search and accumulation is covert, the destination, by association, problematic. Travel to Sydney in late February and early March of every year instantly suggests the SGLMG, age and sexuality being completely irrelevant – it is a time of the year that is, thanks to media coverage, all but completely associated with a particular event.

Accumulation of mental images about vacation experiences is, in the case of SGLMG, almost entirely unproblematic, although the move in this instance has been one from invisibility and oppression to high visibility via media. Like most mental images change is inevitable, but iconic destination images can be remarkably constant. Sydney's traditional iconic attachments are the Harbour Bridge and the Opera House – now, after more than twenty years, Sydney is also the SGLMG, and the iconography now has a new constancy all of its own.

Modification of those images by further information is an annual process for the active seeker of information, or even the unsuspecting viewer who discovers flesh and fantasy as they change TV channels one early March evening. No longer does the seeker have to rely on coy references in brochures – www.mardigras.com.au provides it all, and other links as well. This stage is about refining, or focusing, or shaping – the *organic*, free-form image of the first phase has now become an *induced*, more structured image.

A *decision to take a vacation trip* to Mardi Gras is value-laden, sexuality notwithstanding, and possibly one that may present many challenges for the risk-averse. Depending on the intensity of the desire, the decision to go may have been made long before the actual travel becomes a reality; to make the trip may provide confirmation, or lead to conclusions, that the traveller and those of their contexts – family, education, social, workplace, religion – may not necessarily wish to take a public form. However: 'Because of social disapproval of homosexuality many gay men are forced to find gay space...it is necessary to travel in order to enter that space' (Hughes 1997: 6).

Travel to the destination can take numerous forms that range from the formalized agency/airline package to the individual devising informal, less public forms, depending on the extent to which the travel, and the destination, can be acknowledged.

Participation at the destination can also be problematic when the potential participant may not have an understanding of the operational procedures. While it might seem that participation at Mardi Gras would be intuitive, Mardi Gras as an experience can be so much more than the parade despite its primary media focus. For the gay tourist, parties, recoveries, and the Festival, not to mention the wider Sydney gay community, are all on offer but here individual motivation factors contribute to the decision making process. The extent to which individuals engage, allow themselves to engage, or believe that they are allowed to engage in the event can also be problematic, just as attendance alone can suggest a sympathy or sexual inclination that can also be profoundly problematic if made public.

To *return home* after the event is to potentially enter an aftermath situation that requires careful management. Nothing is the same, and to publicly return from Mardi Gras could exacerbate a situation that had previously been unconfirmed. The return home must acknowledge that a shift has taken place, both in the traveller and their perceptions of the destination, leading to the final phase of *modification of images based on the vacation experience*. It seems very likely that 'as a result of visiting the destination, images tend to be more realistic, complex, and differentiated' (Ross 1994: 76).

And it's over for another year.

The media packages the SGLMG, particularly the parade, into a manageable visual commodity that now is accessed and consumed via print, television and the Internet. It is reported in metropolitan daily newspapers, is broadcast on public access television in a prime-time viewing slot, and has been the subject of documentaries, academic research and many volumes of photography. All of this creates for the SGLMG and its parade a much wider audience than only those who attend the parade, with one consequence being, via the media, its gradual transformation into the biggest cultural and sub-cultural event on the Australian calendar, as well as a significant tourist destination.

With regard to claims about the SGLMG's transformative status, it (being the parade) is still a protest for equal rights and the end of discrimination for those who constitute the Other. The parade is transgression but authorized transgression, breaking the rules but with a dispensation to do so, a fact that the media uses most effectively to continue achieving its desired outcomes.

The parade has grown, just as awareness and, possibly, tolerance of the Other has also grown, at least in early March in Sydney each year. The media/ Mardi Gras (parade) interdependency and interrelationship is firmly in place, each serving, and using, the other. There appears to be widespread evidence, taking many media forms, of a desire for the healthy continuance of such a state, recent financial challenges notwithstanding.

References

Altman, D. (1971; Revised edition 1974) *Homosexual: Oppression and Liberation*, London: Allen Lane.
—— (1997) 'Mardi Gras hits middle age', *The Weekend Australian*, Features 3, Melbourne: March 1–2.
—— (2001) *Global Sex*, Crow's Nest, NSW: Allen &Unwin.
Baker, S. (1966) *The Australian Language,* Melbourne: Sun Books.
Carbery, G. (1995) *A history of the Sydney Gay and Lesbian Mardi Gras*, Parkville, Victoria: Australian Gay and Lesbian Archives Inc.
Clark, A. (2001) 'Dancing in the Street', *Voyeur* (Virgin Blue Magazine), McMahon's Point, NSW: Pacific Client Publishing.
Eco, U. (1984) 'The Frames of the Comic Freedom', in Sebeok, T.A. (ed) *Carnival!* Berlin: Mouton, 1984: 1–9, cited in Seebohm, 1994.
Fakeye, P. and Crompton, J. (1991) 'Image Differences Between Prospective, First-Time, and Repeat Visitors to the Lower Rio Grande Valley', *Journal of Travel Research*, 3: 2.
Fyfe, M. (2001) 'One night in heaven', *The Sunday Age*, News 4, Melbourne, March 4.
Hood, D. (2001) 'Festive fun for everyone', *Sunday Herald Sun,* TV extra 9, Melbourne, March 4.
Haire, B. (2001) 'Mardi Gras'. In C. Johnston and P. van Reyk (eds) *Queer City: Gay and Lesbian Politics in Sydney.* Annandale, NSW: Pluto Press.
Horwitz, T. (2000) 'Fun and Games', *Santa Barbara News-Press*, A9, Santa Barbara, California, September 9.
Hughes, H. (1997) 'Holidays and homosexual identity', *Tourism Management*, 18: 1, 3–7.
Hughes, R. (2000) *Beyond the Fatal Shore*. An Oxford Film production for BBC in association with the Australian Broadcasting Commission/WNET New York and NVC ARTS.
Hurley, M. (2001) 'Sydney', in C. Johnston and P. van Reyk (eds) *Queer City: Gay and Lesbian Politics in Sydney*, Annandale, NSW: Pluto Press.
Johnston, C. (1999) *A Sydney gaze: the making of gay liberation*, Sydney: Schiltron Press.
Johnston, L. (2001) '(Other) bodies and tourism studies', *Annals of Tourism Research*, 28: 1, 180–201.
Knopp, L. (1995) 'Sexuality and Urban Space: a framework for analysis', in D. Bell and G. Valentine (eds) *Mapping Desire: geographies of sexualities*, London: Routledge.
Markwell, K. (1998) 'Playing queer: Leisure in the lives of gay men', in D. Rowe and G. Lawrence (eds) *Tourism, Leisure, Sport: Critical Perspectives*, Rydalmere, NSW: Hodder Education.
Nicoll, F. (2001) *From Diggers to Drag Queens: Configurations of Australian National Identity*, Annandale, NSW: Pluto Press.

Oppermann, M. (1999) 'Sex Tourism', *Annals of Tourism Research*, 26: 2, 251–266.

Overington, C. (2001) 'The race is on for the pink dollar', *The Age*, NewsExtra 3, Melbourne, July 28.

Pearce, P. L., Morrison, A. M. and Rutledge, J.(1998) *Tourism – Bridges across continents*, Sydney: McGraw-Hill.

Powell, S. (2001) 'Shake-out for gay press', *The Australian,* media section 14–15. Melbourne, March 8–14.

Ross, G. (1994) *The Psychology of Tourism*, Melbourne: Hospitality Press.

Ryan, C. (2001) 'Sex Tourism', in N. Douglas, N. Douglas and R. Derrett. (eds) *Special Interest Tourism*, Queensland: John Wiley and Sons Australia Ltd.

Ryan, C. and Hall, C.M. (2001) *Sex Tourism*, London: Routledge.

Scahill, A. (2001) 'Queer(ed) media', in C. Johnston and P. van Reyk (eds) *Queer City: Gay and Lesbian Politics in Sydney*, Annandale, NSW: Pluto Press.

Seebohm, K. (1994) 'The Nature and Meaning of the Sydney Mardi Gras in a Landscape of Inscribed Social Relations', in R. Aldrich (ed.) *Gay Perspectives II: More Essays in Australian Gay Culture*, Department of Economic History with The Australian Centre for Gay and Lesbian Research: University of Sydney.

Swarbrooke, J. (2002) *The Development and Management of Visitor Attractions* (2nd edn) Oxford: Butterworth-Heinemann.

Urry, J. (2002) *The Tourist Gaze* (2nd edn) London: Sage.

Willett, G. (2000) *Living out Loud: A History of Gay and Lesbian Activism in Australia*, St. Leonards, NSW: Allen & Unwin.

Yang, W. (1999) 'Allen and Peter, 1981', in R. Wherrett (ed.) *Mardi Gras True Stories*, Ringwood, Australia:Viking/Penguin.

4 Amber Films, documentary and encounters

David Crouch and Richard Grassick.

Introduction

This chapter is concerned with how media representations of tourism or, rather, tourist spaces work, through a critical attention to recent conceptual developments in terms of the tourist's encounter with space and the possibility of critical, 'alternative' media discourses. A focal point in this chapter is the photographic and ethnographic-type investigation and representation of particular spaces, inhabitants and tourists, in the form of the photo-documentary *People of the Hills* (Grassick and Crouch 1999).

Since the early 1970s, Amber Associates have documented the lives and landscapes of the working class of the North East of England. From a modest ambition to produce small documentary film and photography projects, the group has grown into a feature film production house generating cinema and television dramas, distributed worldwide. Amber have repeatedly been criticised by development agencies, tourist boards and others for generating the 'wrong' image of the region – allotments not coffee shops, quarries not car factories. The defence has been its close relationship with the individuals and communities, arguing that the issue is authenticity and sensitivity to local needs and perspectives, not imposed visions of economic renewal. *People of the Hills,* a photo-documentary work exploring the lives and places of the Teesdale and Weardale communities in the North Pennines confronts this dilemma directly. *Hills* questions the validity of visual representation in tourism materials, exploring the relationships between visitor and local perceptions of the area. Examining the effect of other Amber projects on how outsiders see the North of England, our research considers whether the tourist industry can embrace the working methods of documentary photographers in its materials.

We seek here to engage critically with a radical documentary genre that often works in opposition to dominant ways of mediating places, and consider their potential for visitor engagement. Rather than allowing travel writers to define authenticity of experience for tourists, even in opposition to dominant constructions, this chapter considers the documentary process to be more complex, nuanced, potentially empathetic, but necessarily collaborative. Thus the documentary photographer can contribute to, rather than pre-figure, a process of self-definition and self-realisation. Such a process may include the creation of meaning by those living within particular 'depicted' locations, and similar processes in which the tourist may encounter spaces in acts of embodied practice and performance.

Our convergent cultures are the circulations of mediated material for tourists, communities of individuals living in visited places, and the tourists themselves. We maintain that there is an interesting, potentially productive confluence of ideas between critics of tourist industry-made imagery in representing place to visitors, and the resurgence in the human tradition, here exemplified in the work of arts funded documentary photography. In particular we consider the work of Amber Films and Side Gallery photographers, and our attempts to develop representations of County Durham for the visitor. Side and Amber are situated in north-east England, whose regional culture is significant in this narrative.

Thus we relate two conceptual strands. One concerns the critical, contested debates on relevance in documentary photography practices. The other, a debate surrounding the active encounters tourists may make with mediated stories. It begins with a narrative of the approach and experience of the media collective Amber Films and its experience of documentary work. This story is located in the recent debates concerning the work and place of documentary photography. The narrative shifts to a critical consideration of the idea of the tourist encounter (Crouch 1999). The encounter, which includes a tourist's individual initiative when visiting a place to negotiate what they do in relation to numerous influences, momentarily and over multiple temporalities, may refigure their sense of the world that *Hills* seeks to connect. Whilst Amber's work characteristically does not prioritise addressing the tourist, much of it concerns the mediation of local cultural identities in distinctive places producing representations whose significance may extend beyond these communities alone. We posit a potential openness of the more familiarly polarised tourist of the market perspective; the individual in the visited community and the diversity of mediated impulses drawn through the encounter the tourist makes. Identity and tourist practices pervade this discourse through the idea of practical knowledge and ontology in the way that Clifford identifies as 'cross-cutting determinations not as homelands but as sites of worldly travel: difficult encounters and occasions for dialogue' (1997: 12).

We draw together these orientations by focusing on the possibilities and potentials of photography and film as means of contributing to the process of self-realisation for the tourist as much as for the depicted local community. Further, we engage the challenge of including representations of tourists in a documentary for tourists. This dialogue between ideas of documentary photography and tourist encounters is developed through an experimental documentary produced by ourselves. Crucially, our documentary approach discusses works critically in relation to dominant media constructions of both what cultural and place-identity may be, and of how tourists encounter them. What emerges from this debate is a critical engagement with the role of media production in processes of self-realisation and self-definition, through and between the encounter that tourists and local individuals make with these places, cultures, their own lives, and each other.

Amber and the documentary debate

To consider the 'validity' of this or that image of a place is usually set aside in the interest of the client's wishes, even though a client from one context – say economic development – may demand images of a place which directly contradict those from another – say tourism.

Amber's collective, formed in 1968, sought to work collectively, to produce documents of life in northern England. Side Gallery was opened by Amber, on Newcastle's Quayside in 1977. Both have been long-term recipients of regional arts revenue funding. Side developed strong creative relationships with leading documentary photographers in the UK, Europe and the USA, consistently benefiting from high levels of goodwill and respect in the field. In 1979, Cartier-Bresson celebrated his 70th birthday with a retrospective at the gallery, while his partner Martine Franck was working on a Side commission. Contemporary international work such as Graciela Iturbide's *Juchitan*, Eugene Richard's *Below the Line*, Gilles Peres' *Northern Ireland* or Wendy Ewald's *Portraits and Dreams* sit alongside the projects, commissions and purchases documenting northern England from the seventies: Sirkka-Liisa Konttinen's *Byker*, Graham Smith's *South Bank*, John Davies' *Durham Coalfield*, Chris Killip's *Seacoal*.

Amber's work has also benefited from creative relationships between the interests of photographers and film-makers. The feature films *In Fading Light* and *Dream On* led to and out of many photographic projects in North Shields: the fishing industry, the Meadow Well Estate, a dance school, a local pub. Recently, Amber has concentrated its feature film (*Eden Valley*, *The Scar* and the recently released *Like Father*) and photographic (*People of the Hills*, the *International Documentary Workshops* in Crook) production in County Durham. A new wave of photography commissions explores contemporary lives and landscapes in the post-industrial context of the Durham Coalfield. Ten projects are in different stages of development involving eight photographers and a community documentary group in the ex-colliery village of Craghead. Subjects include single fathers in Seaham, ex-miners and their families, the pollution of *The Coal Coast*, a working men's club in Horden, a small agricultural mart, the South West Durham coalfield, children's views of their lives and futures in a small coalfield village.

This long-running process of image production has resulted in a considerable legacy for the region. Films have been screened at film festivals around the world, in some cases gaining limited theatrical distribution, and broadcast in over 20 countries worldwide to audiences running well into the tens of millions. Amber's latest feature film has won critical acclaim, raising interesting arguments in the region's press, who have contrasted its message of painful personal progress for the region's people to Billy Elliot's totem of personal success leading inexorably out of the region.

Amber's work argues for long-term commitment to communities, encouraging active participation in the production process by those being filmed or photographed, not an easy process. Even in contexts where democratic structures exist, as with some trades unions, there has not always been institutional support for such localised editorial control of work. Whilst Amber built a relatively successful regional magazine with a trade union (Northern News and Views, NUPE 1979 to 1994) involving dialogue with union members, the official structure of the union discontinued the model. Where such structures are absent, as in *Hills*, consultation and participation develops piecemeal. Another Amber inheritance is significant. Robert Flaherty's films were considered partnerships between the film-maker and the people being filmed. *Nanook of the North* was developed on location to enable Nanook to view and criticize rushes, to project the work to the Eskimos during production to 'accept and understand what I was doing and work together with me as partners'. [Flaherty 1920] The photography of the Worker Photographer clubs of the 1930s were a model of participation that later inspired initiatives such as *Casework,* London 1970s.

The experience of sustaining this practice in the North East of England has been recognition and acceptance locally, in the North East and more widely, with an increased marginalisation by the art photography establishment and critique by theorists and commentators in debates about ownership and power, discourses and identification. Side was criticised in the late 1980s in terms of the photographer's right to represent other individuals or communities. New York critic Richard Woodward wrote in 1989:

> This documentary tradition, which not long ago seemed one of the main stems of the photography tree, has of late undergone a crisis of confidence. Ignored by an art world that favors big pictures and high prices, and outstripped by technology that gives the edge to immediacy and narrative scope to videotape, documentary photography has also come under harsh scrutiny from post-modern critics who question its tendency to separate and marginalize groups of people, serving up the work as exotic fare for comfortable voyeurs.
>
> (Woodward 1989)

Reflecting the apparently scientific and objective nature of its technology, photography simply offered a 'mirror on the world'. This widely-held belief had been exploited by photographers struggling to communicate 'their' version of reality to a mass audience less informed than the avant-garde. A typical criticism of documentary photography appeared in European Photography in 1981:

> Document and representation, mirror and window, reflection of reality, these are the recurring terms that appear in the theories which focus on the visual similitude, the analogous moments of a photograph to its objects. It is as if the job of the photographer consists only in the photographic reproduction of an opinion which the photographed objects reflect. It is as if everything ... were already predetermined in the existence of the objects.
>
> (Reinhard Matz: 1981)

Documentarists were told to reveal their own existence in the production process. This was most easily achieved by revealing the author of the image as photographers deliberately cast their own shadows into the image itself. Photographer Jo Spence turned the camera on herself in the quest for a morally justifiable photographic art. The national photography gallery circuit sought acceptance from the national visual arts establishment, and found that more funding potentially became available if photographers took on board the concerns of postmodern art. So Side and its propensity for 'grainy black and white images, mainly an archive' was quietly forgotten. Yet clearly this growth in photography's national standing was its own undoing. The presentation of work grew ever more remote from its production, as galleries vied for an international stage for 'their' photographers. There was little or no time for an ongoing dialogue between photographers and photographed. The original criticism became a self-fulfilling prophecy. Interestingly, demands for a new documentary were made by critics like Allan Sekula and Martha Rosier twenty years ago, calling for a role in social change for documentary work (1978, 1983). Their work, addressed the art world, was decipherable by the avant-garde art world; their subjects were passive outsiders.

Central to this discussion is the contemporary conceptual documentary photography debate, which Side Gallery has consistently championed. Photographers in the documentary tradition continued to address humanitarian debates, not artistic qualities of their working methods. The arts funding world, where support is based on a work's ability to generate an artistic rather than socially driven debate, prevailed. The debate about content was overtaken by an obsession with modernism and form, then by a cynicism of 'engagement'.

Amber argued that its work was indeed subjective, and that the artist/photographer had a role to play in interpreting the world. The phrase 'the creative interpretation of reality' came to be widely used to describe the abilities of the various photographers commissioned by Side to document the North East of England. Side continued to argue for a different kind of creative space in which to operate, first opening a 'current affairs' photo-essay exhibition space in 1989, producing a policy document arguing for a series of five community-based photographer activists, creatively to interpret the communities around them, producing their own work while organising local workshops/exhibitions; Side's status as the region's focus for photography devolved to five smaller projects, working closely with local communities, whilst the city centre gallery co-ordinated these projects.

The response was dramatic. The region's arts association withdrew its support for Side, notoriously stating verbally that the model represented 'double-arms-length-funding', and could not be justified.

Instead, Northern Arts requested proposals for a new Photography Gallery to explore in its programme the use of photography as artform rather than as documentary. Side closed for a year, and several Amber members did not find project support, one having difficulty photographing people since her early 1990s study of mothers and daughters in a North Shields dancing school was attacked by feminists at a gallery in Edinburgh. After a lull in funding, the region now acknowledges documentary work again, giving Amber fuller funding.

These debates concern two interconnected issues: the sharing of authority/control over work; the language employed in that work. Photographers anxious to gain recognition on the international art stage may have given up on any dialogue with the subject of their work – and indeed have stopped photographing 'subjects' and ridicule the simplistic language of documentary imagery. Documentary media workers continue to value advantages of working closely and long-term with communities. The overriding issue hindering this is economic survival. Photographers have historically tackled this by self-support, organising co-operatively.

The documentary photography tradition offers a fruitful basis for shifting the use of photographs away from the re-presentation of tourism product to a more socially engaged role. In the flux of debates one community-based project was established in County Durham. Grassick developed a circuit of touring exhibition venues, thus *People of the Hills*. Challenging earlier sign-driven explanations of tourism new perspectives focus on how the *tourist* encounters space and its stories. From this perspective we inform an interpretation of the potential of *Hills* to participate in a cultural convergence.

Tourist encounters and place/culture mediations

Space, landscape, heritage and other content often regarded as 'static' remain powerful in the way in which places are inured in tourist seduction by contemporary capitalism – the business sector and its marketing (Selwyn 1996). Yet it may be that the continuing prioritisation of mediated images and their consideration *a priori* in tourism analysis is misplaced (Mordue 2001). Photography is essentially a visual medium that contributes to the wider resource of mediated images in text and pictures. For Lash and Urry, this 'making use' of signs 'out there' became a focus for the mental reflexivity of the subject's own life in a practice of postmodern subjectivities (1994). Following Baudrillard's desire this suggests that people come to respond to their life through the provided spectoral seduction. However, Lash and Urry argue a subject-centred and less prefigured or dependent seduction, or even a reverie of self-seduction as an escape in an everyday sense of wanting change, adjustment, rediscovery, of emotion, identity and relations. In contradistinction to a 'floating signs' perspective, they claim that human subjects live their lives, live space, encounter each other, work through their own lives in a negotiation of the world around them/us.

It is, then, appropriate to posit another way of looking at the contextualisations of tourism/the tourist experience, and knowledge derived therefrom. Several strands of recent debate deviate from the traditional orienting of the discourse of what tourists do from prefigured meanings and instead focus on the tourist and consider how they engage available mediations; that is, think in terms of the working of tourism mediations. To do this we outline recent reconceptualizations of the tourist, in terms of the available mediated images in a more complex flow of encounter through which those signs may be deconstructed and engineered into an agenda constituted by the tourist. Such an approach enables a profound reconfiguring of how mediated images work, and are worked by the tourist. We suggest that there are multiple arrivals and departures through which diverse mediated materials are encountered, and perhaps, temporally, made sense of.

We maintain that desire rests at least in part in people's feelings and actions, conveyed, figured and developed in their embodied self, mind/body, body/mind, inadequately presentable in the language (Crouch 2001). Such a perspective is oriented we argue, to the increasingly humanistic postmodernism emerging from across the disciplines (Crouch 2001, 2004; Crouch and Lubbren 2003), and more to a sympathetic, though not uncritical, subject, in Levinas' terms, unfolding, negotiating and contesting rather than constantly finding the world always unstable, empty of identity and overwhelming (1961). We problematise the privileging of vision accorded in these critiques of tourism and its modernisms. Indeed it would seem to be over-reductive to interpret tourism as sight-seeing. The moments of sightseeing on any tourist experience are fleeting, not necessarily essential (in terms of meaning) and should be understood in terms of a much wider, more complex practice.

Photography is notably problematic in how it relates to what individuals do. Foucault described how features, content and their relationship are encountered (our term) in chaotic ways (Philo 1992). Do tourists set up their cameras where they do merely as an expression of the pre-programmed tourist? It may be because these points are convenient and practical or mark an achievement they have made. Yet the value given to the photographs taken may be contextualised just as much from their own feeling at the time, their company and so on (Crang 1997). Recent work on material culture argues that objects can be given new significance through the ways in which they are consumed. Consumption is understood as an active process whereby objects, products, places, and things can be made *to matter* (Miller 1998). Sites, destinations and particular locations are constantly being refigured. Programmed visual culture of tourism can be similarly refigured, as is evident in diverse versions of London Bridge in Arizona (Jewesbury 2002). To practise and to perform are components of the flow of contemporary culture. Meanings change and are changed.

Indeed, as Matless argues, representations are themselves the temporal product of practice (1999). The production of other artwork has been unpacked in terms of an explorative sensing and seduction in the encounter with place and space (Crouch and Toogood 1999). A more relational discourse is needed between the tourist, the image-producer and the visual objects, and of how such objects may be given significance.

Recent components of so-called 'non-representational' theory have developed insights into what tourists do and how they make sense of the world from embodied practice and performance. Its emphasis is on the individual as mind/body construction, not merely thinking, simultaneously bodily engaged in the world. The individual not only thinks but also does, moves and engages the body practically and thereby imaginatively, and in relation to material objects, spaces and other people. The individual is surrounded by spaces rather than acting as an onlooker. Crucially individuals' encounters with the world include diverse cultural contexts. Yet these contexts are not determining, and their significance is modulated and negotiated through practice (Nash 2000).

Merleau-Ponty built on extensive empirical research to argue that individuals are engulfed by space which they encounter, both multi-sensually and multi-dimensionally and are thereby likely to be similarly informed. Touch, a feeling of surrounding space, sight, smell, hearing and taste are worked interactively. All the senses are involved. However this mode of embodied practice does not operate as a gathering device but is worked through the way the individual uses his/her body expressively – it turns, touches, feels, moves on, dwells. Rather than set aside the individual is engulfed by the space around him/her. Cloke, P. and Perkins, H.C. (1998) exploration of the individual acting bodily in their discussion of white water rafting contend that the individual as embodied human being engages the world through feeling, as connector as well as receiver. Things, artefacts, views and surrounding spaces become signified through how the individual feels about them.

Furthermore the character of the surrounding world is not merely gathered in data but is engaged, and constituted expressively (Radley 1995). There is a power of being poetic through which places, events and things are given value. Soile Veijola and Eva Jokinen explored the enactment of the bodily sense through a discourse on the body and the beach. Their work is formulated both through the appearance of the body as inscribed in conventions, expectations and projected desire, and in terms of the bodily practice of the beach by sensuous, and sensual, and poetic beings (1994). Individuals can overflow the boundaries of the rational and the objective and be playful, imaginative going beyond what is evidently 'there' in an outward, rational sense. De Certeau described this capture of space as a poetics of practice, as Birkeland explored in her discussion of the experience of the North Cape in Norway (de Certeau 1984: Birkeland 1999). Furthermore what is done is frequently done in relation to other people, i.e. intersubjectively, and the character of events, experiences and sites is being additionally attributed through what people do in relation to others (Crossley 1995). For tourists this may include both other tourists and people who are at that moment not tourists, but locals and so on (Crouch 2001).

The social anthropologist Tim Ingold has used ideas of embodied practice in thinking through how individuals may thereby relate to the world through the practical things they do and how they do them, that Harre has termed 'the feeling of doing' (Ingold 2000, Harre 1993). Individuals visiting a place may encounter it performatively. Performance concerns the enaction of life through protocols of engagement, surrounded by ritualised practice, working to pre-given codes habitually repeated, conservative, working to cultural givens. Yet it can also be potentially disruptive and unsettling, or at least have the potential of openness in refiguring space and the self in relation to those protocols (Carlsen 1996). The general discourse on performance, informed by performance studies, understands the individual's actions to be done in relation to the self or to others, 'performed for', including self-regulation and negotiation intersubjectively. Practice and performance colour the character of consumption. Visual material and prior narratives can be disrupted in this process, or may not be as significant in the desire of the encounter, or its enactment.

As Shotter argues from a social constructionist perspective in his explo-ration of knowing and feelings, the individual works the practice of everyday life ontologically, making sense through what is done and how it is done (Shotter 1993). And Burkitt has engaged with the embodied character of this process of practical ontology and its emergent ontological knowledge and its potential role in refiguring the self and negotiating identity (1999). Embodied practice is engaged in the flow of individual reflexive thinking (Lash and Urry 1994) to constitute the character of their knowledge. In part at least individuals produce their own geographical knowledge through what they do and think (Crouch 2001), in relation to contexts and representations, not in detachment from them.

These radical adjustments in how the tourist is understood enable us to rethink, critique and reconstruct our understanding of how the media works in the practice of being a tourist, in doing tourism. Such a perspective enables us to rethink the work of producing media images and texts in conversation with, rather than merely targeted at, the tourist, and in collabo-rative production with local communities in a documentary tradition.

People of the Hills: making documentaries: communicating with tourists: the tourist encounters.

In our ten-year partnership our concern has been to be critically reflexive in the ways in which we handle this material. We do not seek to promote this particular method, but to explore how recent theoretical developments on the tourist encounter and critiques of documentary photography may be engaged. The challenge is to find ways of working that are true to the humanist traditions of documentary photography, and engage both photographer/photographed in meaningful and important debate about our lives.

People of the Hills involves producing and exhibiting hundreds of images of Teesdale and Weardale communities. Bringing these images back to the people appearing in them, we were struck by the importance local people attached to '*their*' view of their area. Important details in the landscape, experiences, memories, all shaped their view of what is important about their place – and lives. Many of the images and notes were exhibited in local towns and villages, where the subjects of the images and families and friends could attend. They have participated in the developing discourse concerning the process and context of making the images; what they have said has been fed back into the making of images of local life, surrounding spaces, individuals who constitute both, our interpretations of them, a kind of detail usually missed by representations for the tourist constructed by the tourism industry.

Hills was motivated in part by the dominant media images of the Northumberland and Durham Tourist Board, whose narratives of 'heritage', of history and power are captured in their discourse of 'Land of the Prince Bishops', denoting the historic realm of power in the land as the key construction of what this place means, of how it may be given value. The main content of this representation is one of grand views, overwhelming land, ownership, yet something that is somehow both accessible and shared by the tourist. These corporate presentations usually omit local, contemporary voices, projecting an image, an idea, an identity that exercises a dominant culture and projects a context through which the tourist will make their visit. Go and see it; feel part of it even; centuries-old heritage fixed, awaiting the visitor. The place 'offered' is conditioned as a 'tourist product', constructed as pre-packaged, ready in situ thus proscribing the visitor tourist experience.

Hills offers an alternative representation of this place and how its heritage is constituted and understood. Richard Grassick sought to make images of the contemporary behaviour of individuals living in the area, and of the places they encountered in their lives. Several slices of *People of the Hills* used in the eventual work were made, including women working in a local factory, allotment keepers and smallholders, children at school, and quarrymen. The working process was to spend time getting to know these individuals exploring the spaces they used, in work and elsewhere. Through informal

Figure 4.1 Nelsons' and sky

conversations built up over sometimes months of intermittent visits he established friendships with them, taking his family on invited return visits.
Narratives of their lives were constructed from ethnographic-type structured
interviews, from which were developed portfolios of images. Interviews by
both authors took place at times between and around the photography, local
exhibitions providing the time and space for further feedback from subjects.

The interim stage of *People of the Hills* was published in 1999. The next
phase of this work is to build comparative collections with visitors, already
including caravaners. Over 80 photographs, 10,000 words across 84 pages,
text and photographs work together to combine visual images and the
subjects' narratives of their lives, community, valleys and hills where they
live. The intention is that the voices speak, in their narratives, albeit always
through an edit of ours (quoted voices, below, from these narratives).

The photographs combine composed, large views, some romantic and
dramatic: 'I just love that view, I could sit here and look at it all day'
(Interviewee Elaine), and close up, fragmented images of people moving,
pushed by the wind and doing work; celebrating a moment in their lives, and
so on.

I love it up here, I just love these hills. I just love that view. I could sit
and look at it all day. It changes so much you can sit and look at it for

Figure 4.2 Image of Old Quarry

an hour and it changes in that hour. Even in bad weather it is still a spectacular view. I would not like to live down in the valley. It is just so open up here.

This documentation seeks not to romanticise their lives, nor trivialise them, but to say something through conversations and photo-responses of the individuals concerned.

Observations can be both romanticising and brutally realistic: 'Fetching water to unfreeze the pipes. The more remote farms can go for weeks without running water'. Inter-cut, close-up and fragmented images seek to bring the encounter with the tourist closer, possibly making it easier to include a glimpse of the content of the cultural worlds that 'work' this landscape into their encounter, yet not as curios, because their images are self-selected, and accompanied by their voices, often hard edged and uncertain.

I remember, ten years back there was the prospect of a crusher mate and I thought this might be a decent career move because in those days I thought mineral extraction would be the thing in the dales. I saw something in the press with these power stations needing x million tonnes of limestone and thought these old quarries round here would be opened up and they'd need staff for that. But that doesn't seem to have come about.

Figure 4.3 Caravanner outside van

Some of the voices engage in a debate with tourists, their expressions expectedly varied: '... here in the bed and breakfast we find some of them are reluctant to go for a walk in case they trespass ... we're doing a little bit to [get] closer together' (Interviewee Harold).

Visitors' voices do not relay *Land of the Prince Bishops*, but tell their own stories:

> In the caravan, you open the door and you go out with what you've been in bed with; you open the door and nobody says anything. It's the natural thing to do isn't it, it's just the open air space – it's what you've come from, people just expect it. You don't have to wear a suit or anything; you don't have to wear any particular type of clothes, you just come and you're caravanners and that's it!

As this illustration expresses, people give character to the hills in their own time-space (all quotations in this section derive from *Hills*). We have added our own text to the interviews section of the volume, attempting to engage with the visitor, which may be intrusive, or enabling, in what the visitor does in their encounter anyway. In a 'chapter' headed *A Good Walk*;

> Making knowledge of a place like this is both a kaleidoscope and patina. The place surrounds you, and this is felt whether in working and taking the children to the leisure centre or in wandering a footpath or just walking around the car visiting the Dale for the first time. (1999: 76)

We argue that this combined narrative – of the subjects, and 'editors', affords the capacity of such documentation to communicate the character of this place as constructed and constituted by the individuals. Of course there is selectivity. The manifest selectivity we argue, is a part of what documentary photography seeks to achieve. The narrative as a whole is a subjective one, and thus not different from that provided by the Tourist Board. However *Hills* communicates a particular reality, a different *reality* [sic]. Rather than deliver another product, or facet of that product, *Hills* offers an open discussion, however selectively, in which the tourist can make an encounter, or series of encounters. The voices encounter the tourist; the images sometimes look out to the tourist too, but sometimes demonstrating their enjoyment in play than as in the course of getting on with their lives. When tourists are depicted, as in the case of caravanners, they may look out to the reader/viewer and speak through their text. We argue that the voices do not project continuous stories of 'fun places to be' but experiences, lives, curiosity, on their own terms. Stories invite the tourist into a debate about what makes the culture work, and the multiple self-identification of the individuals themselves. The interim book is intended to be available as a series of booklets, possibly also in a version mediated by the Tourist Board, to visitors. Its work offers the reader/visitor a flexible story to be used in different ways, with different entries and experiences.

Can this material engage in what the tourist is doing? We try and disrupt the idea of the places and the individuals' lives as merely the object of the tourist gaze, even in trying to go beyond Urry's terms of an anthropological gaze (2000). The subjects we talked with and photographed were happy for their stories and images to be used openly. What we are doing seeks to enter what tourists are doing anyway, not claiming to start the encounter process, but to engage with it. However, centrally, we seek to offer an alternative way of encountering through a distinct, questioning, somewhat awkward, perhaps more challenging documentary. The variation and incompleteness of the images may offer easier access to the multiple discourses of this place, rather than providing a 'set-piece'; multiple opportunities for thinking about how culture and place works; fluid lives, spaces to be encountered, to enter into where the tourist is, what the tourist is doing. Although, it may be premature and presumptuous to claim that this kind of representation becomes more effective in collaboration and partnership with the tourist, it could offer an open and malleable contribution to the tourist's own practice of self-realisation and self-identification, and contribute too to the similar process in which these local workers and residents are involved.

Tourism industry media for the tourist increasingly offer a more open text. After listing sites and events they note 'do whatever you enjoy most' in several variations of text. What *Hills* seeks to achieve is a combination of more multiple texts and available identifiers, but it self-consciously does so in an effort to convey a distinctive professional documentarian's approach, which approach is of course widely open to contestation. It does not claim to be the 'right' story, but another story. The tourist is then given a choice *and* is opened to a series of possibilities of ideas, of routes, both material and metaphorical.

Hills offers a narrative of heritage. So does the Tourist Board. Their approaches are different. For *Hills* the heritage mediated is one that is contemporary; an encounter with the world that is multiple, worked in practice; that engages with historical character and contemporary negotiations. But this representation is still a 'product', and its influence in offering alternative ways of seeing or not is affected by other available materials and their marketed competition, or by any particular ways in which the prepared media of *Hills* is delivered to the tourist.

Clearly the eventual images are not the individuals' images, but professional ones. It attempts more an open narrative than a presentation of the tourism product. Yet in so doing it may emerge as merely another wrapping of a tourism product, and certainly the possibilities of its alternative narrative may be influenced by the ways in which it is made available to visitors. Individual tourists may use the work as a means to 'fix' a curious and strange (sub)culture, or use it to develop their encounter with the world into something different. Further, if tourists are increasingly reflexive, even as post-tourists, they may value the opportunity to engage in their encounter with what other tourists do and feel. This narrative seeks to contribute to the dialogue in which tourists engage.

Documentary image making – a valid representation of the region for visitors?

Hills represents an attempt to intervene in a discourse of local identification and to convey that to visitors, but it may not succeed in either offering more than product facets or features, or as offering any more accessible contribution to the tourist encounter. It may do so. An evaluation remains to be done, but this is common to so much media work. It does convey fragments of individuals' lives to themselves and contributes to a process of self-identification and self-realisation. It makes available stories otherwise out of reach to the visitor, seeking to do this in a 'colourful' way (rather than only as serious documentary). The authors' narrative is intended to address the tourist directly in what they may be doing, feeling, touching, seeing and thinking as they use their tourist experience to negotiate their life. This could result in a translation from the self-identification of the local people and what they feel their lives and place to be in the process of seeking to convey something to visitors.

The ideas behind this project have sought to relate, rather than combine, an approach to documentary photography and its role in the self-realisation of a series of local populations, and a perspective on the tourist encounter and its potential to be experienced in a process of self-realisation and self-identification. Both practices emerge as multiple, open and fluid. Although the visitor and the 'local' encounter the place through very different contexts and practices, there are elements of encounter with space that suggest there are overlaps between their encounters rather than a distinct separation.

Our idea of documentary photography eschews the particular postmodern claim that its method is naive, determined by a particular version of objects essentially 'there'. We acknowledge the selectivity of the professional process but advocate the opening of multiple stories and diversity of self-definition, one part of which is in terms of ideas of heritage, as fluid, discursive and practised, repeatedly made and remade. The tourist can enter, or encounter, that process through this narrative. Moreover, heritage is accessible, or can be made accessible in the documentary process. Furthermore, that heritage as process is less concerned with authenticity for the local individuals than about ownership, a feeling of control over a version of what the place is. For the tourist, this can provide a version of heritage that connects more than may be apparent. As noted above, the work of cultural mediators is often regarded as delivering authentic images that presume construction of the tourist experience by tourism and other media. Such interpretations usually prioritise the presumed desire of the tourist for authenticity, and that it is delivered through the auspices of mediation. If we concur that places are encountered in multiple routes, then the visitor him/herself participates in a process of figuring and refiguring heritage, rather than of passively consuming prepared and prefigured heritage as product. However we have sought to construct a critical means through which the operation of professional cultural mediation can be made sense, whilst acknowledging that *Hills* offers a particular narrative, constructed through collaboration between professionals and other individuals, through/alongside which other versions of heritage may be considered or contested.

In summary we proffer a theoretical debate on the working of documentary photography through recent insights on doing tourism. This becomes problematised in terms of the ways in which the texts of the mediators are encountered, confronted and consumed by tourists. Rather than concentrate upon how the media may inform, or be resisted by, tourism media narratives, we argue that it is appropriate to rethink how the media and tourist encounters work by interrogating how they *relate*, through processes of self-negotiation in the encounters individuals, as tourists and as locals, make.

We have presented here some of the criticisms of documentary photography practice that have grown out of a shift in the context for such work from socio-political and humanist concerns to commodified art. Many concerned photographers still wish to contribute to social change and progress through their work. The project *People of the Hills* is one attempt to locate a project in a context other than photography as art, or indeed journalism. Instead, it seeks to position itself in a discourse in the context of a local and wider debate about the tourist industry's marketing priorities, thus enabling all kinds of local participations to come into play.

References

Baudrillard, J. (1981) *For a Critique of the Economy of the Sign Telo*, St Louis

Birkeland, I. (1999) 'The mytho-poetic in Northern Travel', in D. Crouch (ed.) *leisure/tourism geographies*, London: Routledge.

Burkitt, I (1999) *Bodies of Thought: Embodiment, Identity and Modernity*, London: Sage.

Carlsen, M. (1996) *Performance, a critical introduction*, London: Routledge.

Clifford, J. 1997 *Routes: Travel and Translation in the late twentieth century*, Cambridge, Massachussetts: Harvard University Press.

Cloke, P. and Perkins, H.C. (1998) 'Cracking the canyon with the awesome foursome', *Environment and Planning D: Society and Space,* 16:185–218

Crang, M. (1997) 'Picturing Practices: research through the tourist gaze', *Progress in Human Geography* 21 (3) 359–373.

Crossley, N. (1995) 'Merleau-Ponty, the elusory body and carnal sociology', *Body and Society* 1:43–61

Crouch, D. (1999) *Leisure/Tourism Geographies*, London: Routledge.

—— (2001) 'Spatialities and the Feeling of Doing', *Social and Cultural Geography* 2 (1): 61–75.

—— (2005) 'Flirting with Space', in C. Cartier and A. Lew, *Seductions of place*, London: Routledge.

Crouch, D. and Lubbren, N. (2003) *Visual Culture and Tourism*, Oxford: Berg.

Crouch, D. and Toogood, M. (1999) 'Everyday Abstraction in the Art of Peter Lanyon', *Ecumene* 6, 1: 72–89.

de Certeau, M. (1984) *The Practice of Everyday Life*, Berkeley: University of California Press.

Grassick, R. and Crouch, D. (1999) *People of the Hills*, Newcastle: Amber/Side Gallery.

Harré, R. (1993) *The Discursive Mind*, Cambridge: Polity Books.

Inglis, F. (2000) *The Delicious History of the Holiday*, London: Routledge.

Ingold, T. (2000) *The Perception of the Environment: essays in livelihood, dwelling and skill*, London: Routledge.

Jewesbury, D. (2002)'Tourist: pioneer: hybrid, the mirage in the Arizona Desert', in D. Crouch and N. Lubbren (eds) (op. cit.) 223–240.

Lash, S. and Urry, J. (1994) *The Economies of Space*, London: Sage.

Levinas, E. (1961) *Totality and Infinity* (trans. A. Lingis) Pittsburgh: Duquesne University Press.

MacCannell, D. (1999) *The Tourist* (first edition 1976), Berkely, University of California Press.

Matless D. (1999) *Landscapes of Englishness*, London: Reaktion.

Matz, Reinhard *Against a Naive Concept of Documentary Photography*, European Photography, Spring 1981 issue

Miller, D. (1998) *Material Culture: why some things matter*, London: Routledge.

Mordue, T. (2002) 'Heartbeat Country', *Tourist Studies* 2, 1.

Nash, C. (2000) 'Performativity in practice: some recent work in cultural geography', *Progress in Human Geography* 24 (4): 653–664.

Philo, C. (1992) 'Foucault's Geography', Environment and Planning D: Society and Space Vol. 10, pp. 137–61.

Radley, A (1995) 'The Elusory Body and Social Constructionist Theory', *Body and Society* 1 (2) 3–23.

Rosier, M. (1983) 'Some Documentary Photography', *Afterimage*. 11 (1–2): 13–15.

Sekula, A. (1978) 'Dismantling Modernism. Reinventing Documentary (Notes on the Politics of Representation)', *Massachusetts Review* 19 (4): 859–883.

Selwyn, T. (1996) *The Tourist Gaze*, Chichester: Wiley.

Shotter, J. (1993) *The Cultural Politics of Everyday Life: social constructionism, rhetoric and knowledge of the third kind*, Buckingham: Open University Press.

Spence, Jo, (1985) *Putting Myself in the Picture: A Political, Personal and Photographic Autobiography*, London: Camden Press.

Thrift, N. (1997) 'The Still Point: resistance, expressive embodiment and dance', in S. Pile and M. Keith (eds) *Geographies of Resistance*, London: Routledge, pp.124–54.

Urry, J. (1995) *Consuming Places*, London: Routledge.

—— (2000) *The Tourist Gaze*, London: Sage (2nd edn).

Veijola, S. and Jokinen, E. (1994) 'The Body in Tourism', *Theory, Culture and Society* 11:125–151

Woodward, Richard B. [1989] *Serving up the Poor as Exotic Fare for Voyeurs?* New York Times 18 June 1989.

Filmography

Nanook of the North, d. Flaherty, Robert J., 1922

5 On the actual street

Nick Couldry

Introduction

What can the study of the media contribute to the study of tourism? On the face of it, electronic media at least (which is what we principally have in mind when we talk in common-sense terms about 'the media') change the organisation of space by making available a 'despatialised' awareness (Thompson 1993: 187) of other places. Electronic media might seem, therefore, to make actual journeys across space less important. Instead, however, I will be arguing that media representations of the social world make certain places more important, reconfiguring the landscape within which tourism occurs. New 'compulsions of proximity' (Boden and Molotch 1994) undermine generalisations about the supposed evacuation of space and place in postmodernity, and media tourist sites are a good example of such compulsion.

Interest in the media's impacts on the wider landscape of consumption has been gaining momentum for some time, and geography, as a discipline, has been central here. If an earlier generation of geographers (Meinig 1979: 183; Jackson, 1994: viii) lamented the loss of a symbolic landscape based in architecture and place, recent work has explored how media references have helped create a new symbolic landscape. The 'magic' of mediated place encompasses shopping malls (Kowinski 1985; Hopkins 1990; Langman 1992) and theme parks, particularly those which are sites of current or historical media production (Davis 1996; Gottdiener 1997; Couldry, 2000: Part Two). Sharon Zukin captures a more general trend in the changing interrelations between place and media when she claims that DisneyWorld's architecture matters 'not because it is a symbol of capitalism, but because it is the capital of symbolism' (Zukin 1991: 232). Rather than reduce this new landscape to an extension of the audience's supposed passivity before television (Sack 1992, chapter 5), it is more useful to attend to its details, and the divisions and hierarchies that structure them; in short, to take seriously the idea that this landscape is a 'landscape of power' (as Zukin puts it), with all the complexity that implies.

This chapter will discuss material from a detailed study I made in 1995–98 of visits to the outdoor set of Britain's longest-running soap opera,

Coronation Street, which is housed at Granada Studios Tour, Manchester ('GST') on a site next to the Granada Television studios. I will move outwards from more straightforward aspects of why people visit GST to more adventurous suggestions about the ritual quality of 'the Street' set (as it is often called) as a pilgrimage site.

Studying the set of Coronation Street

The set of *Coronation Street* (GST's principal attraction for many, perhaps a majority of, visitors) is a place of paradox. Its visitors pay to visit a location they have already watched free on television for years: part of the pleasure is not seeing something different, but confirming that the set is the same as something already seen. The Street set undoubtedly has a 'power of place' in Dolores Hayden's (1995) term, and yet, on the face of it, is poorly qualified to satisfy Hayden's definition of the term (ibid: 9): 'the power of ordinary urban landscapes to nurture citizens' public memory, to encompass shared time in the form of shared territory'. The Street set is, of course, only an image of an 'ordinary urban landscape': no one has ever lived or died there. Clearly, the Street set's 'power of place' rests not on public history in the usual sense, but on shared fiction. It is, as we shall see, a place with 'aura', a 'ritual place'. I want to explore the framework within which visits to GST are meaningful. This is not to ignore issues of economics (marketing strategies, leisure resources, and so on). On the contrary, the high cost of visiting GST (both money and time) makes it all the more important to establish the meaning of the place which attracts such expenditure.

What do people do on the set of *Coronation Street*? They walk down it: people sometimes summed up their visit in this phrase. But, since many spend an hour or more on the set, there must be more to the visit than that. People take photographs and are photographed at points of interest – outside the Rovers Return pub, the shops, the houses – but that too is over quickly. Almost everyone spends time testing the boundaries of the set's illusion: looking through the houses' letter boxes or windows, pressing doorbells and knocking on doors; looking round the houses' backs (the 'old' houses have paved yards backing onto an alley, the 'new' houses have gardens). People compare the details of the set with their previous image of the Street, testing, for example, if the set is up to date with the plot. Some of the set's details are aimed at visitors, not the television audience: for example, the 'for sale' notices in the newsagent's window. There is a lot of laughter on the set, especially when it is crowded. There is, of course, the pleasure of pretending, for a moment, that you live on the Street, posing with door knocker in hand or calling up to one of the characters. The visit is an elaborate form of performance and exploration.

A significant minority of visitors will already have visited GST before. Of the 21 people who wrote to me in the course of my research, six had visited more than once and another seven said they wanted to return. Of the 143 people I interviewed on site, 21 per cent had visited before, some more than once. (I also interviewed 11 people off-site in 9 interviews, usually in people's homes; it is these longer interviews which are drawn on most in what follows; first names of off-site interviewees have been changed for reasons of confidentiality.) Taking the site sample, returnees were three times as likely to be women as men; they also (as one would expect, given costs) were more likely to live in GST's own region (76 per cent from the North of England). Even if returnees are a distinct subset of visitors, the fact that people return to GST *at all* needs to be explained. Chris Rojek has written of 'the sense of anticlimax that often accompanies the visit' to contemporary tourist sites: 'we see it; but have we not seen it before in countless artifacts, images, dramatic treatments, and other reproductions?' (1993: 196). The risk of anti-climax would seem to be especially great at GST (every visitor has seen the Street countless times on television). Not only, however, were such comments rare, but the routine nature of some people's visits may depend precisely on the Street set's *taken-for granted* symbolic significance: 'it's something everyone does and that's it. It's like ... the Tower of London ... I mean, you wouldn't sit down and discuss your visit to the Tower of London with people'. Perhaps visiting the Street set is significant precisely because it is the place you routinely watch. We need to unravel the implications of this apparently simple claim.

Most people I interviewed were positive about their visit to GST. There is, of course, a wide spectrum of engagement. People may visit out of interest: to see 'what goes on', 'how it all works', finding it 'educational'. There is the pleasure of participation in the fiction, seeing '*Coronation Street* come to life'. But the visit may also involve considerable emotional investment for both men and women. For John, the intensity of going to the Street was 'like being on a drug'. Some people said they found it difficult to believe that they were actually there – on the set. Underlying all these reactions is the sense that it is significant to 'be there': it is an 'experience' marked off from 'the ordinary'. As one man put it: 'I want to see the place ... where this thing is, you know. It's an absolute experience, isn't it, a magnificent experience, isn't it, to come to this place.' Being on the Street set, then, is intrinsically significant.

How can we go beyond this starting point? As mentioned, it is the shared framework of significance underlying people's visits to the Street that I want to explore, a framework which may be shared both by those bored and those fascinated with it. That does not mean, however, that it is just the most common reactions that I discuss. I will put considerable weight on those most intensely engaged with the set, not because their detailed reactions are necessarily typical of the wider sample, but because, by putting so much weight upon the shared framework of significance, they reveal its thought-patterns most clearly.

Questions of identity

A first step is to consider visits to the Street set as public expressions of identity. One obvious significance is as an affirmation of northern English, working-class identity, for which the programme *Coronation Street* has provided a widely recognised stereotype for almost forty years (Dyer et al. 1981; Geraghty 1991; Shields 1991: 222–229). The dangers and constraints of this stereotype were occasionally remarked upon critically by 'Northerners' who had been to the South or explicitly reproduced by interviewees from the South . The stereotype is, of course, only partly negative; the associations of a 'Northern' sense of 'community' are positive. A connection with their own living conditions was acknowledged by visitors who were themselves Northern and/or working class and talked of *Coronation Street* as just 'ordinary', 'everyday living'. However *Coronation Street*'s image of (Northern) working-class life has been rejected as outdated by other soaps, whether representing the 'South' (*Eastenders*) or the 'North' (*Brookside*) (Geraghty 1991: 34).

The question, then, of what *Coronation Street* represents – and therefore what visiting GST, and the Street set, might signify – is already a complex one. For Beth, *Coronation Street* 'is our heritage … our culture', a sign of not just Northernness but also of Englishness, like 'the smell of green grass'. Issues of identity connect with the rhetoric of GST and Granada Television (companies with a mission to 'represent' the region), as this comment of an HGV driver from Lancashire made clear:

> Being in the North-West, it's on your doorstep, I've watched it for years … I've been brought up watching it … and here we are [laughs]. You know, it comes up on telly and you think, Ah, it's only round the corner that, bit like your local … Like the tour rep said, it's put us on the map sort of thing …

The overall position is, however, more complex. Not only does *Coronation Street* represent only one in a whole field of competing representations of Northern and/or working-class identity on British television. Also, for many, it is strongly associated, not with social reality now, but with the past: whether a personal past ('a breath of home') or, more starkly, a social past that is lost. As a middle-aged couple from Warrington put it:

MAN: [sighs] I think [CS] is a place that no longer exists in reality really.
WOMAN: They have tried to update it but comparing it to where we live that neighbourly spirit has disappeared, you know.
MAN: Yeh, and mainly due to television … people come home from work or wherever and they shut the front door and they switch on the telly and that's the end of it.

The irony that the community which *Coronation Street* projects has been destroyed in part by television itself is powerful.

In any case, visiting GST must mean more than a simple affirmation of class or regional identity. For any affirmation of identity at GST is complicated by the fact that the programme is, as everyone knows, a fiction and at GST you see how that fiction is constructed. That may bring disappointments quite separate from any wider sense of social identity affirmed by being there. For Beth, as mentioned, the programme's connection with her life (particularly her childhood) was intense. But, reflecting on her visit, she felt disappointment:

> It's like when you were a child, you imagine something, then you go back to it as an adult and it's totally different ... it was exactly like that. Everything just seemed very small and flat.

Issues of identity are here cut across by issues of fiction. Indeed, it is striking how little people spoke to me *explicitly* about class or regional identity. Perhaps it was too obvious to mention; certainly the dynamics of the interview situation (with me being a middle-class Southerner) may have encouraged some reticence. But another important feature of GST as a tourist site is that there is no place where class and regional identity are explicitly focused on *as such*. Many visitors to GST, in any case, are neither working-class nor from the North. There is, however, another identity affirmed at GST which cuts across regional and class divides: the 'community' of the programme's fans. A number of people mentioned their pleasure in the 'camaraderie' on the site that this may generate. It was expressed by Susan and Glenys (both lower- middle-class Southerners):

SUSAN: ... as you're looking over there, you say something, and then somebody behind you will say, Oh, so-and-so and so-and-so. And you tend to get ... into another crowd then ...

GLENYS: So we were all there ... with a common ... thought, that we wanted to see Corrie ... so you could talk to people and know that you had something in common even if it was only the fact that we were all *Coronation Street* fans.

The community of fans connects people who do not know each other, across different regions and classes; visiting the set may crystallise a temporary sense of that community, what the French sociologist Michel Maffesoli (1996: 11, 13) has called the 'empathetic "sociality"' we feel when we find 'those who think and feel as we do'. Once again, however, it is dangerous to generalise: from my observations, most people went round the set on their own or with the small groups in which they came. There was certainly a sense of sociality on the set – expressed most outwardly in laughter – but visitors generally experienced this *in parallel to, rather than with*, each other.

To understand GST better as a tourist site, we need, then, to go beyond the obvious shared identities performed at the site, and look more closely at the detailed ways in which visiting the set makes sense as a practice.

Being on the Street

The basis for the Street set's significance is, seemingly, very simple: it is the place where the programme's filming goes on, the actual place you have watched from your home over the years. There is of course an important fantasy element to being on the Street, the feeling you are in the place where the cast are filmed: it is 'quite magical really, to actually believe that you're there on the spot where ... the stars walk along' (Barbara). But this imaginative connection with the programme's fictional frame depends on fact: the fact that the set *is* the place of external filming. What are the implications of this?

The Street's significance as the place where filming in fact goes on was marked routinely in people's language. To be on the Street set is to be on the 'actual Street', to 'be there' at the place where 'programmes are actually made'. Its houses are the 'real' places of filming, not mere 'studio sets' or 'mock-ups'. This is a principal reason why people go to the Street set: 'I just wanted to see where it was done' (Julie); 'good to see the actual street where the show is filmed'. The fact that the set is the actual place of filming was not something most people quickly registered before moving on. I often heard people testing it out, wanting it confirmed:

WOMAN (20S) IN LARGE GROUP OF WOMEN ASKS GUIDE: 'Do they really film here?'

MAN (MIDDLE-AGED) ASKS THE GUIDE: 'Is this the original *Coronation Street*?' 'No', he's told, 'this was built in 1982'. 'Where was the original one?' 'It's where the New York Street is' is the answer. 'But it was all done here [i.e. on this site]?' 'Yes', says the guide. (extracts from Fieldnotes)

Not everyone was convinced of this fact. There were some people who rejected the idea that the Street set is significant because it is a place of filming. Such counter-opinions emerged occasionally when I interviewed a couple or larger group: for example, the view that historical tourist sites are more 'authentic' than media tourist sites. These, however, remained minority voices among those I met – not surprisingly, since my sample was weighted towards those interested in the Street set.

Visiting the Street set may even involve an element of dislocation. If television 'constantly invokes ... an unmediated experience that is forever absent, just beyond a hand reaching for the television dial' (Anderson 1994: 82–83), then collapsing this distance may be experienced as puzzling: 'it's really weird though walking on it, because you watch it on TV and then you're thinking, well, people actually walk down this Street filming'. The sense of strangeness may continue when you reflect back on the visit much later:

DEBBIE: I don't know, it's sort of like being in a dream really, thinking I'm actually walking down *Coronation Street*. I just couldn't believe it ... it just doesn't seem real sometimes.

For some people, the significance of 'being there' – on the Street set – goes beyond what they can rationally explain. For John, there was a sense, almost, of privilege:

I know that's silly because literally millions of people go a year now, and millions of people have seen it, but I felt that I was the only one, I felt I was there and I'd seen it for so long, and ... it was like a dream come true, really.

A Canadian woman, originally from India, visiting during a holiday in Europe, put it this way:

It's hard to express what I felt when I walked up to the Street to actually feel I was there, I mean I think that's going to stay with me for ever. Because it was such a wonderful feeling, it just left me speechless, you know, I just wanted to stand there.

Why does it matter so much 'just to stand there'? And why does it matter so much to show others you have been there?

GLENYS: That is what we were there for, wasn't it, to see the Street ... We've got a photograph ... on our wall in our room at work, of the two of us outside the Rovers [Susan laughs] ... and that's us [laughs]. And we've got lines all round it, so everyone can see it. We've been there [laughs].

All media-based tourist sites, of course, involve a sense of 'being there' ('there', the place from which a media narrative has been generated). As with many clichés, however, it is a mistake to dismiss it too quickly before examining the pattern of thinking condensed within it. This will provide us with a better understanding of how media-based tourism works as a symbolic practice.

Aura

We saw how important it is for most visitors to know that the Street set is the actual place of filming. This relates to a distinctive feature of British soaps: what Christine Geraghty has called their 'regional authority' that comes from representing a place with a regional identity (1991: 35). An extension of this notion of authenticity is the assumption that the soaps are filmed in real places, or at least that they are produced in a place situated in the narrative's region. Here is one man from Lancashire who had emigrated to Canada:

> Anybody watching any show in the States, if they went to Miami, I mean they may not see people from *Miami Vice* because it may be filmed somewhere else entirely, but we know this [CS] is filmed here, you see.

We also have seen that, for some visitors, the Street is a place with a precise history, associated with specific episodes (Barbara). Another woman wanted to enter the set of the Rovers Return because 'there's a lot things happened there over the years'. In each case, the Street set is regarded as a place with a history which is 'fictional' only in a general sense (the sense in which *Coronation Street* as a whole is a fiction). If we regard the Street set as a place of filming (the perspective most people adopted), the set has a real history – of filming – tied to the history of the programme's narrative. It is the *real*, not fictional, place where fictional events were actually filmed.

That sense of history was at issue when John rejected going to Blackpool to see Granada's 'World of *Coronation Street*' exhibition.

> I've no desire to go, I would hate it, because it's not the real one. [short laugh] All right, so people could say, 'But that [CS] isn't the real one'. But it is, it's where they film the outdoor scene, it's the one where the actors are, where the studios are, where it all originated. Where did Blackpool come into it?

Others made similar comments. One woman who was visiting Manchester during a holiday in Blackpool put it: 'no, we thought we'd come to the original'. The Blackpool 'Experience' is 'not the real Street'; only the Street set itself is 'the real thing', 'the real place'. Why? Because 'you know it's all done here'.

Note that John's sense that the Street set, and only the Street set, is worth visiting exists despite his knowing that it is 'only' a set and that others regard it as such ('all right, so people could say, "But that isn't the real one"'). This might seem to confirm the 'postmodern' truth, wittily expressed by Umberto Eco, that we live in a world in which the 'completely real' is identified with the 'completely fake' (Eco 1986: 7, quoted in Rojek 1993: 160); the Street set from one point of view is (as John understands) only 'fake'. But, again, to leave our analysis there would be a mistake.

People's preferences for seeing the 'real Street' (the 'original') are interestingly at odds with Walter Benjamin's famous thesis on the loss of 'aura' 'in the age of mechanical reproduction' (Benjamin 1968). What people who reject the Blackpool 'Experience' hope to obtain at GST is precisely an 'aura'. Not the 'aura' of something outside the 'mechanical reproduction' of filming, but the 'aura' of the place and process *of filming itself*: using Benjamin's phrase, 'its unique existence at the place where it happens to be' (1968: 220), the aspect of a place that can only be grasped by going there. As Debbie put it in relation to the Street set, 'people never appreciate it, unless they're there'. Returning to John, 'aura' for him is not just some general notion of 'being there' inherent in any media site, but a quality precisely tied to the set's material history. Benjamin defined 'authenticity' in just this way: 'the essence of all that is transmissible from [the object's] beginning, ranging from its substantive duration to its testimony to the history which it has experienced'(1968: 221). Compare that with John's explanation of why the Street set is better than a mere studio set:

> I have seen studios ... but nothing to compare with the Street ... When you're sitting in the studio, you do see ... the unreality, but on the Street ... it's a real street, albeit there's nothing behind the door as such. But you're still there, it's still real. ... There was a funny thought that went through my mind, that it had been raining. ... And I actually looked down and thought, this is real because there's real rain, it sounded so stupid. And I stood in a puddle and I thought, Oh Crikey! Yeh, this is real, it's not covered over, it's always outdoors ... the actors go out in all weather ... it's real rain and it's real cobbles and it's real dirt [laughs] ... You don't expect a set to be that real.

John was not the only person to regard the rain on the set as significant. Why is the rain so special for John? Hardly just because it confirms the physicality of the set: even a studio set is physical in this respect. The rain is significant in part, I suggest, because in a small way it is 'a testimony to the history' which the set 'has experienced' (Benjamin); it is a token of the set's authenticity and John's authentic experience of it, his definitive access to its 'aura'. The rain which has fallen, and will remain, on the set after John has gone allows him to project into both past and future the connection – between viewer and Street – that 'being there' involves.

The Street set as pilgrimage site

I want now to develop this notion of connection by exploring a ritual dimension to what people do on the Street set. Here is how one multiple visitor, Michael, described being on the Street set in a letter to me:

> From the moment I put my foot on the Street I feel like a star. I start my walk down the Street starting from the 'Rovers Return' to the 'corner shop'. I look through *all* the windows and through *all* letter boxes. I touch the stone cladding of number 9. I feel so so very happy and trouble free when I walk down the Street ... I just can't believe it. Every

time I walk down the Street I get that same wonderful happy feeling. ...
It [GST] is the best thing and most wonderful thing I have ever done.

There is a palpable sense of ritual here. Again, rather than dismiss it as
eccentric, we should contextualise it in terms of what is perhaps the Street
set's most fundamental attraction: its status as 'ritual place'.

In order to explain that claim, we must return to the basic question: what
(for all visitors, not just devoted fans) does being on the Street set involve?
Being on the Street involves a comparison between what you have watched
over the years and the set itself. On the face of it, this is a banal comparison
(seeing if the Street 'is actually like it is on telly'), but its dimensions are
worth considering.

First, you are linking things in two different time-frames, the years during
which you have watched the Street and the time now when you walk onto
the set: 'for me, it was amazing because I've seen it on the TV for so many
years now. ... For me it was brilliant to finally see everything'. It is the
bringing together of two separate time-frames (the time of your regular
watching over the years, the time of your visit now) that allows a sense of
completion: 'to finally see everything'. In Barbara's account, the transition
between the time-frames of long-term watching and present visit is reproduced exactly in the transition from the final video image of the Street to the
sight of the Street set itself:

> You went in a room where they showed you a video of sort of past
> episodes, and then they drew the curtain back. You'd watched it on the
> telly and then it was actually there. And then you set off and then you
> walked along it.

The feeling of walking into the space of the screen itself is vivid: this
'freedom' is clearly part of the designed effect of sites such as GST (cf Rojek
1993; Davis 1996). It works, however, partly because it reproduces in miniature the transition between time-frames that being on the Street itself
involves. That is why, for Barbara, there was 'no point of actually going on
the Street and then doing the video'. Second, being on the Street involves
comparing the results of two different activities, two ways of looking, for
which sometimes people used different words. 'Watching' the Street on television, you are constrained in how you can look at the set: you are limited by
camera-angles, and so on. 'Seeing' the Street set close up allows you to look
at its details in your own time and from any angle, and then put the whole
thing back together:

> I spent quite a bit of time there [on the set] and then after lunch I went
> back there and took a small turn ... you know standing back and seeing
> it and picturing it in my mind as to how it appears on TV.

Seeing what the set 'is actually like' is an active process of finding out, qualitatively different from watching television. There is also a third, spatial, dimension to the comparison. 'Watching' the Street is something we do in the home, whereas 'seeing' the Street set can only be done in GST's public space. Being on the set therefore connects two normally separate sites of discourse: the home and the site of media production. All these dimensions (time, activity, and space) are combined in Julie's comment:

> It was nice to see. An experience that you ... actually sit in your living room and you're actually watching that place, but now you're actually standing in that place, and you can say ... I've actually been there, and it felt good.

'Being there' involves connecting your 'everyday' practice of *private* viewing with the *public* place where the programme is actually filmed. This connection of different times, places, and activities is neither neutral nor trivial:

> It's magic, it's a great feeling, sitting at home when you watch telly and say I WAS THERE! To think you could do that.

> Just nice to know that you've seen [it], when you watch telly, that you've actually been and seen it for yourself.

As the last two quotations suggest, the significance of *having* been there goes on being enhanced after the visit is complete. This is because the connection made by 'being there' can be relived when at home you watch the programme again:

> [Susan:] I mean, we were *unbearable* when we first came home, because as soon as it came on, [we said] We were there! [Glenys laughing]. And that's where we stood! ... Every time we see it, we think, [whispers] Oh we've been there! [Glenys laughs] And it's still, it's still there, Oh, we've been there. It's really good, you know.

Since the private/public connection made by 'being there' on the set is intrinsically significant, just the basic acts of occupying space on the Street are significant in themselves: 'to actually stand in the Street is lovely' (woman); 'just walking up and down something you see regularly in front of your eyes' (man). It is enough that you are, or were, 'there'.

All tourist sites involve the realisation of some private/ public connection: a visit is always preceded by a private act of expectation. There is much more, however, to the Street set's 'power of place' than that. For the force of the connection the set embodies is the way it formalises and spatialises the hierarchical relationship between the 'media world' and the non-media, or 'ordinary', world. Not only is this hierarchical relationship significant and pervasive in contemporary societies (a large claim, of course, that I do not have space to defend here, but see Couldry 2000), but it is precisely the type of category hierarchy that is played upon in ritual practice. The work of the anthropologist Jonathan Smith (1987) is helpful here. He has drawn on Durkheim and Levi-Strauss's accounts of symbolic classification to develop an original account of ritual place. 'Ritual' he argues:

> relies for its power on the fact that it is concerned with quite ordinary activities placed within an extraordinary setting.... Ritual is a relationship of difference between 'nows' – the now of everyday life and the now of ritual place; the simultaneity, but not the coexistence, of 'here' and 'there'. Here (in the world) blood is a major source of impurity; there (in ritual space) blood removes impurity. Here (in the world) water is the central agent by which impurity is transmitted; there (in ritual) washing with water carries away impurity. Neither the blood nor the water has changed; what has changed is their location. This absolute discrepancy invites thought, but cannot be thought away. (1987: 109–110)

On the Street set, analogously, people do ordinary things – walking up and down, looking in shop windows, and so on – but they do them in an extraordinary setting (the frame of the Street set). Indeed, the whole process of being on the Street, as just argued, brings out connections – and differences – between the 'ordinary' process of television viewing (the 'now' of everyday viewing) and the 'extraordinary' moment of the visit (the 'now' of being on the 'actual Street'). The two situations remain of course separate, and the difference 'cannot be thought away': it is a difference within a symbolic hierarchy. The set is not any space, any street, but the 'actual Street' that you and everyone else have been watching all those years from your home. It is, in this precise sense, a ritual place, where two 'worlds' are connected.

The ritual dimensions of visiting the Street set encourage us to take our analysis one conceptual stage further, and see these visits as, effectively, 'pilgrimages'. The metaphor of 'pilgrimage' has become so routine, so laden with irony and parody, that it has, arguably, lost analytic value – if, that is, we regard clichés as empty. I prefer however to follow the social psychologist Michael Billig's argument that it is precisely the patterns of *banal* language that, by attrition, reinforce large-scale patterns of thought which are anything but banal in their consequences (Billig 1995; 1997).

The general significance of the 'pilgrimage' cliché beloved of journalists and also academics (Reader and Walter 1993) – a chosen journey to a significant place – derives from the way that contemporary societies are overlain, but unevenly, with shared narratives of significance. The landscapes of contemporary tourism are an important way in which such narratives are enacted and embodied. We make 'pilgrimages' to distant places which have not only personal significance, but a guaranteed social importance too; they matter to an imaginable group of others, even if when I set off on a pilgrimage I do not know who in particular I will meet on that journey. 'Pilgrimage' points are potential gathering-points where the highly abstract nature of contemporary social connection can be redeemed, through an encounter with a specific place. In general sociological terms, therefore, pilgrimage points are places where the 'disembedded' nature of late modern communities can be 're-embedded' (Giddens 1990) in the form of a journey to a chosen, but distant site.

'Pilgrimage' in this broad sociological sense, far from being a trivial aspect of the modern social world, is endemic within it; contemporary tourism (the commerical organisation of significant exceptional journeys) is saturated with possibilities of 'pilgrimage'. Media pilgrimages are specifically journeys to points with significance in media narratives. Through media pilgrimages, not only is the abstract nature of the media production system 're-embedded' in an encounter, for example, with a site of filming or a celebrity, but the significance of places 'in' the media is more generally confirmed. The media pilgrimage is both a real journey across space, and an acting out in space of the constructed 'distance' between 'ordinary world' and 'media world'.

To use the word 'pilgrimage', however, is not to claim any religious significance for such media-related journeys. In line with Durkheim's general rethinking of religious experience in terms of experiences of sociality, one leading concept of pilgrimage (Victor Turner's) encompasses many journeys without any link to religion:

> Both for individuals and for groups, some form of deliberate travel to a far place intimately associated with the deepest, most cherished axiomatic values of the traveler seems to be a sort of 'cultural universal'. If it is not religiously sanctioned, counseled or encouraged, it will take other forms. (Turner and Turner 1978: 241).

So there is no 'sacrilege' in extending the term pilgrimage to secular forms, including tourism. On the contrary, the faded religiosity of this term captures exactly the sense of continuity that we need to grasp the condensed resonances of contemporary media-saturated tourism.

GST is a 'pilgrimage' point in the sense that it is a central, symbolically significant place, where 'special' time can be spent apart from the time of 'ordinary' life (cf. Turner 1974), time that is 'special' simply because spent within 'media space': 'your time on the Street'. What is affirmed by going there is not necessarily the specific values (if any) associated with *Coronation Street* the programme, or even with the act of watching it. What is affirmed, more fundamentally, is the 'value' condensed in the symbolic hierarchy of the media frame itself: its symbolic division of the social world into two incompatible parts, a non-media world (where we live) and a media world to which we may (exceptionally) travel.

Conclusion

Dean MacCannell's pioneering 1970s research into the social and cultural resonances of tourism, for all its theoretical richness, received less attention than it deserved, perhaps because of the totalising, functionalist, neo-Durkheimian framework that it implied (although MacCannell himself tried to move beyond that framework: 1992: chapter 11). MacCannell rightly rejected Daniel Boorstin's reductive (1961) analysis of all tourism as just travel to 'pseudo-places' already seen on television, and offered instead the more constructive interpretation of tourism as a 'form of ritual respect for society' (1973: 589) and the huge range of work that complex contemporary societies contain. The links between MacCannell's vision and the 1990s growth of tourism to places where the media work (film or TV locations) are now clear, even if, in making them, we need to re-emphasise the power dimensions of the symbolic landscape within which such journeys are meaningful (the hierarchical relation between 'media' and 'ordinary' worlds being an aspect of contemporary social power).

From our perspective today, we can see such journeys, and their growing media component, not as postmodern aberrations but as part of a wider intensification of the centralising processes of modernity. Ernesto Laclau in his book *New Reflections on the Revolution of Our Time* captures the paradox at work here very well: '[contemporary societies] are required by their very dynamics to become *increasingly* mythical. This is linked to the proliferation of dislocations peculiar to advanced capitalism … commodification, bureaucratic rationalisation, and … the increasingly complex forms of division of labour' (1990: 67, added emphasis). One of the primary myths of the contemporary world is that 'the media' are our central access-point to whatever we might want to call social 'reality' (cf. Couldry 2003: chapter 3). It is no longer then surprising that occasionally we wish, many of us, to spend our scarce time and money visiting the places where the media productions that instantiate that myth are made. Such sites of media tourism are, after all, not visits to just any place of work. They are visits to the places where the images are produced through which 'society' imagines it sees itself.

Acknowledgement

A more detailed account of my research at Granada Studios Tour is set out in Couldry (2000) Part Two; for a fuller development of the concept of 'media pilgrimages' suggested here, see Couldry (2003) chapter 5. Note that as of the time of writing, I understand that GST is closed for refurbishment, although private visits to the Street set are still possible by special arrangement.

References

Anderson, Christopher (1994) 'Disneyland', in H. Newcomb (ed.) *Television: The Critical View* (5th edn), New York: Oxford University Press, pp. 70–86.

Benjamin, Walter (1968) 'The Work of Art in the Age of Mechanical Reproduction' in *Illuminations*, London: Fontana.

Billig, Michael (1995) *Banal Nationalism*, London: Sage.

——(1997) 'From Codes to Utterances: Cultural Studies, Discourse and Psychology', in Marjorie Ferguson and Peter Golding (eds) *Cultural Studies in Question*, London: Sage, pp. 205–226.

Boden, Deirdre and Molotch, Harvey (1994) 'The Compulsion of Proximity', in R. Friedland and D. Bodern (eds) *NowHere: Space Time and Modernity*, Berkeley: University of California Press, pp. 257–286.

Boorstin, Daniel (1961) *The Image: or whatever happened to the American Dream*, London: Weidenfeld and Nicholson.

Couldry, Nick (2000) *The Place of Media Power: Pilgrims and Witnesses of the Media Age*, London and New York: Routledge.

——(2003) *Media Rituals: A Critical Approach*, London: Routledge.

Davis, Susan (1996) 'The Theme Park: Global Industry and Cultural Form', *Media Culture & Society* 18(3): 399–422.

Dyer, Richard *et al.* (1981) *Coronation Street*, London: BFI.

Eco, Umberto (1986) *Faith in Fakes*, London: Secker and Warburg.

Geraghty, Christine (1991) *Women and Soap Opera*, Cambridge: Polity.

Giddens, Anthony (1990) *The Consequences of Modernity*, Cambridge: Polity.

Gottdiener, Mark (1997) *The Theming of America: Dreams Visions and Commercial Spaces*, Boulder: Westview Press.

Hall, Stuart (1973) 'The "Structured Communication" of Events', Stencilled Occasional Paper No. 5, Birmingham: Centre for Contemporary Cultural Studies.

Hayden, Dolores (1995) *The Power of Place: Urban Landscapes as Public History*, Cambridge, Mass: The MIT Press.

Hopkins, Jeffrey (1990) 'West Edmonton Mall: Landscape of Myths and Elsewhereness', *Canadian Geographer* 34(1): 2–18.

Jackson, John Brinckerhoff (1994) *A Sense of Place a Sense of Time*, New Haven, CT: Yale University Press.

Kowinski, William (1985) *The Malling of America: An Inside Look at the Great Consumer Paradise*, New York: William Morrow.

Laclau, Ernesto (1990) *New Reflections on the Revolution of Our Time*, London: Verso.

Langman, Lauren (1992) 'Neon Cages: Shopping for Subjectivity', in R. Shields (ed.) *Lifestyle Shopping*. London: Routledge.

MacCannell, Dean (1973) 'Staged Authenticity: Arrangements of Social Space in Tourist Settings', *American Journal of Sociology*, 79(3): 589–603.

—— (1992) *Empty Meeting Grounds*, London: Routledge.

Maffesoli, Michel (1996) *The Time of the Tribes*, London: Sage.

Meinig, D.W. (1979) 'Symbolic Landscapes: Some Idealizations of American Communities', in D. W. Meinig (ed.) *The Interpretation of Ordinary Landscapes: Geographical Essays*, Oxford: Oxford University Press.

Reader, Ian and Walter, Tony (eds) (1993) *Pilgrimage and Popular Culture*, London: Macmillan.

Rojek, Chris (1993) *Ways of Escape*, London: Routledge.

Sack, Robert (1992) *Place, Modernity and the Consumer's World: A Relational Framework for Geographical Analysis*, Baltimore, MD: The Johns Hopkins Press.

Shields, Rob (1991) *Places on the Margin*, London: Routledge.

Smith, Jonathan Z. (1987) *To Take Place*, Chicago: University of Chicago Press.

Thompson, John (1993) 'The Theory of the Public Sphere', *Theory, Culture and Society* 10(3): 173–189.

Turner, Victor (1974) *Dramas, Fields and Metaphors*, Ithaca, NY: Cornell University Press.

Turner, Victor and Turner, Edith (1978) *Image and Pilgrimage in Christian Culture*, Oxford : Basil Blackwell.

Zukin, Sharon (1991) *Landscapes of Power: From Detroit to Disney World*, Berkeley: University of California Press.

6 Screaming at The Moptops:
Convergences between tourism and popular music

Sara Cohen

Introduction

On 23 August 1996, four young men took to the stage in a large bar room located in the basement of a grand but rather jaded hotel in Liverpool, a post-industrial maritime city on the northwest coast of England. Each wore a grey round-collar jacket, white shirt, black tie, pointed black patent boots and a black wig. Together they performed a series of Beatles songs on instruments similar to those of the original and infamous Beatles, including Paul McCartney's notorious left-handed bass. Their vocal and instrumental sounds, stage movements and gestures, were closely modelled on those of the original band members, hence 'Ringo' moved his head rhythmically from side to side behind his drumkit which displayed in large black letters the band's name, The Moptops. 'George' kicked up his feet in a familiar quirky, knee-jerk style, and the heads of 'Paul' and 'John' met as they leant into one of the microphones, shaking their dark heavy fringes. The bar contained around two hundred people. Some wandered around the darkened perimeter walls looking rather lost and lonely. Others queued to buy drinks or stood further forward sipping their drinks and chatting in groups of two or three. Towards the centre of the room a large attentive group of mostly young women gathered around the spot-lit stage. Amongst them were those who sang along and danced enthusiastically or screamed at the band, gesticulating wildly at each song. Following the final encore some giggled excitably as they queued backstage to demand autographs, but the band were soon ushered away by the event's organisers.

The Moptops performance is used in this chapter as a starting point for an exploration of convergences between music and tourism.[1] Music and tourism have always converged in some way. Music sounds, scenes and performance events have encouraged people to visit geographical places in person, or to travel to other places in an imaginary sense (Cohen 1998). Music has also been used to represent and market places to potential visitors because of its ability to connect places with particular images and emotions. In addition, music has influenced visitors' experiences of places, because of its central role in local leisure and entertainment venues. This chapter focuses on a specific form of niche tourism based on visits to locations significant to the lives and work of popular musicians, using the case of Beatles tourism in Liverpool as an example. The main aim is to explore the increasing role and significance of cultural tourism[2] in British and North American post-industrial cities during the 1980s and 1990s, and the growing emphasis on, and legitimation of, popular music as part of this process. It examines how this involved convergences between popular music and tourism events, industries and policies, and highlights convergences and similarities between discourses of and about popular music and tourism. The chapter ends by highlighting significant disjunctures between tourism and popular music cultures in order to offer a better perspective on their intertwining.[3]

Tourism and popular music events

Many events could be categorised in terms of music or tourism or a combination of both. They include eighteenth-century classical concerts or operas that attracted elite audiences on the Grand Tour, twentieth-century club nights in Ibiza that have attracted young, pleasure-seeking backpackers (Connell and Gibson 2002: 231–232), Broadway shows and seaside pantomimes used to promote package holidays and weekends, and the Moptops performance which brought together social and occupational groups involved with music and tourism.

The Moptops were a professional Beatles tribute band from Britain. Tribute acts are so called because they model themselves on other well-known performers, attempting to recreate their performances. During the mid-1990s the popularity of tribute acts became widespread in Western Europe, North America and Australia, and have become a familiar feature of tourist attractions based on a few specific popular musicians, most notably Elvis Presley, Dolly Parton and The Beatles. The Moptops' performance was part of the 'Welcome Party' for Beatles Week, an annual celebration of the Beatles staged every August in Liverpool. Beatles Week involved a convention based in a local hotel, aimed at Beatles fans, particularly those from outside Liverpool, and events related to the convention such as the sale and auction of Beatles memorabilia, Beatles-related exhibitions, talks on the Beatles, and numerous performance events and competitions involving visiting professional and amateur Beatles tribute acts from all over the world, from Sweden's Lenny Pane to The Parrots from Japan and The Beat from Argentina.[4] Most tribute musicians, including members of The Moptops, declared themselves to be strong Beatles fans saying that they performed largely out of love and admiration for the Beatles and their music. Many performed alongside tribute acts for other artists at the Mathew Street Festival which marked the official end of Beatles Week, and involved performances by musicians on outdoor stages and in bars and pubs across Liverpool city centre.

The audience for The Moptops included some Beatles fans and musicians from Liverpool, but largely consisted of visiting fans, an extremely diverse group in terms of age, class, gender and nationality, representing varying degrees of interest in the Beatles. There were those whose interest in the Beatles was fairly casual, plus many ardent fans who said that they 'lived and breathed the Beatles every day' planning much of their lives around Beatles-related activities and events. Some had accumulated vast collections of Beatles recordings, artefacts and knowledge, some ran Beatles fanclubs or wrote for Beatles fanzines. Many regularly attended Beatles conventions in Europe and North America but said that the Liverpool convention was particularly significant because Liverpool as the birthplace of the Beatles had influenced and inspired the band and their musical creativity. For some this was their first visit to Liverpool but other fans were referred to as 'returnees' because they returned to the city to attend the convention every year. Fans attended the Liverpool convention to learn more about the Beatles' connections with Liverpool and to celebrate and commemorate those connections. They believed that by visiting the city that had inspired the Beatles they might feel closer to the band or experience something of their 'aura'. Plus, they wanted to meet and interact with other fans to establish and maintain friendships and social networks. The Beatles and their music had attracted visitors to Liverpool since the 1960s. They were perhaps the first band to make their local origins a part of their global success. They connected themselves to Liverpool in interviews, through lyrical compositions, and were associated with a Liverpool scene and sound (Cohen 1994, 2003c; Connell and Gibson 2002: 98).

The Moptops' performance event was organised by Cavern City Tours (CCT) then Liverpool's only city-based tour operator. The company launched in 1983 as a specialist in Beatles tours, and had driven the commercial development of Beatles tourism in Liverpool as its most central player. CCT organised Beatles package weekends and ran a daily 'Magical Mystery' coach tour around Liverpool sites of Beatles significance. In 1986 it took over as the main organiser of Liverpool's annual Beatles convention (this began as an informal event organised by local fans but was later run by the regional county council), an event officially recognised in 1995 as the region's second most important annual tourist attraction. By 1993 CCT had four full-time employees and had staged the first annual Mathew Street Festival. Throughout the 1990s CCT continued to extend Beatle-related tourism and leisure activities. The company took over ownership of the Cavern Club, made famous by the Beatles who performed there on a regular basis between 1961 and 1963, and also launched the Cavern Pub. Both club and pub were located in Mathew Street and CCT planned to develop a Beatles theme hotel and restaurant nearby. By 2002 CCT had 150 employees, including part-time bar staff (Kaijser 2002).

Other commercial organisations directly involved with Beatles tourism in Liverpool at the time of The Moptops performance included the Beatles

Story museum, the Beatles Shop selling Beatles recordings and merchandise, and a few individual entrepreneurs. The directors of CCT referred to those businesses as Liverpool's 'Beatles Industry'. Beatles tourism in Liverpool was thus a small-scale, private-sector affair, although according to the statistics of the regional tourist authorities, in 2002 it contributed £20m to the region's economy. There were also businesses which were more indirectly involved with Beatles tourism, including the hotel that hosted the Beatles convention, and a group of private-sector businesses based in or around Mathew Street, some of which had launched the Cavern Quarter Initiative in 1993. This initiative involved the development of a music heritage quarter around Mathew Street, and aimed to physically transform and gentrify the area through place marketing and the development of tourism, retail and entertainment.

Despite some rivalries between the commercial operators involved with Beatles tourism, they formed a close-knit group drawing upon similar phrases, anecdotes and arguments when they discussed Beatles tourism. Many were Beatles fans driven also by the commercial potential of the Beatles' connection with Liverpool, arguing that for many people Liverpool was synonymous with the Beatles, and that this was something the city should capitalise upon by making The Beatles the focus of organised tourism, and by protecting and preserving city buildings with connections to the Beatles. They categorised city visitors according to their interest in the Beatles and their corresponding economic impact. CCT, for example, distinguished between 'fans' who visited Liverpool on a 'once in a lifetime' Beatles 'pilgrimage', 'fanatics' or 'die-hards' who visited Liverpool repeatedly from all over the world, and 'general interest' visitors who visited the city for other reasons besides The Beatles (Dave Jones, Personal Communication, April, 1994).

The Moptops' performance event thus involved tourist entrepreneurs, musicians and music fans, and was also perceived in terms of both tourism and music. In one sense, for example, the Moptops were a musical tribute to a globally successful pop band, members of their audience comparing their performance with other popular music events involving tribute bands or the Beatles. Simultaneously, however, some audience members also related the event to tourist events experienced during visits to other places, The Moptops being likened to a heritage centre because they were designed to re-enact and recreate the local past through repertoire, stage dress , performance gestures and movements.[5] The Moptops performance event could thus be perceived as involving a musical representation or construction of Liverpool for city visitors, because of the close connection between the Beatles and the city where they were born and brought up. The Beatles may be regarded as a product of Liverpool but the city has also been a product of the Beatles and their music. Liverpool features in several Beatles songs, and the Beatles' wit, their 'scouse' accent and characteristic nasal tones (Belchem 2000), and their jangly guitars and melodic, male vocal harmonies have been used to signify Liverpool in various contexts, to promote and market Liverpool as a tourist destination.

Tourism and music institutions and policies

By bringing together groups and interests, activities and events involved with tourism and music, The Moptops' performance exemplifies the growth of cultural tourism in British and North American cities during the 1980s and 1990s. It also points to increasing convergences between tourism, the arts and other cultural industries, and between cultural and economic policies.

Liverpool has always attracted visitors because of its maritime heritage, grand architecture and waterfront, and its impressive array of museums and galleries. Throughout the twentieth century, however, Liverpool experienced a series of economic slumps far more severe than those experienced by other British cities. During the 1980s and early 1990s, in particular, the city became known in Britain as much for urban decline as for football and The Beatles. The British media used the city as a symbol not just of urban decay but of everything wrong with the nation's cities, representing the city through narrow, ugly stereotypes. Liverpool's economic and image problems contributed to a long-running ambivalence within the city to the idea of promoting The Beatles as a local tourist attraction. According to the director of the Merseyside Tourism and Conference Bureau, there were 'those who feel that Liverpool shouldn't have to rely on four lads – there's more to it than that', and 'those who feel that the Beatles were local lads who made good and then turned their backs on the city' (Pam Wilsher, Personal Communication, 4 October 1993).

As a symbol of rock culture's conventional dreams of escape from the local, the Beatles served to highlight perceptions of Liverpool as a shrinking city in decline. Tourist slogans and banners proclaimed 'From Liverpool to the World', and 'Liverpool – the 5th Beatle' concealed deep, long-running tensions within the city. Additionally, some policy-makers associated The Beatles, and popular music generally, with drugs, deviancy and rebellion. During the 1980s tourism and city re-imaging were not on the agenda of Liverpool City Council which was dominated between 1983 and 1987 by an extreme Trotskyist faction, the Militant Tendency, which prioritised local taxation and housing policies, regarded culture and tourism as peripheral concerns, and entered into a direct confrontation with central government, threatening to bankrupt the city.

Throughout the 1990s CCT and other Beatles tourist entrepreneurs lobbied city policy-makers for financial investment and recognition, arguing that the Beatles were an important resource for the city and a potential tool for its regeneration. They complained of a lack of local support for their important and unacknowledged role in the promotion, marketing and regeneration of Liverpool and in the preservation of its heritage. To strengthen their argument they likened the connection between Liverpool and The

Beatles to that between Stratford and Shakespeare (Dave Jones addressing a meeting of the Merseyside Music Industry Association in 1994), thus elevating The Beatles to the status of high culture. Through promotional literature CCT informed clients, largely Beatles fans, about their ongoing struggle with city policy-makers, but here their emphasis was slightly different, highlighting the value of The Beatles as city heritage and criticising city policy-makers for their unwillingness to preserve and promote that heritage and to pay The Beatles the respect they so obviously deserved. Meanwhile CCT suggested to local music-makers that The Beatles could be used as a 'seed' for the promotion of the city's music more generally. They argued, for example, that the Mathew Street festival need not necessarily be associated with one narrow aspect of Liverpool's musical past but could be broadened out to incorporate other music genres and styles. As part of their efforts to develop Beatles tourism in Liverpool CCT thus operated as intermediaries between policy-makers, music-makers and fans, tailoring the way that they interpreted connections between The Beatles and Liverpool in a manner suited to each group.

Several factors combined, however, to encourage Liverpool policy-makers to pay more attention to Beatles tourism and to cultural tourism more generally. First, following Britain's economic decline during the late 1970s and early 1980s, The Greater London Council had pioneered a new strategy for the regeneration of Britain's cities emphasising the contribution of art and culture to local economic development, that connected, for the first time in Britain, cultural and economic policy. The strategy was supported by high-profile research (see Myerscough 1988) that emphasised the role of art and culture in generating local employment and improving the image of cities, thus helping to attract visitors and investment that would enable them to reassert or increase their influence on 'the world stage' (Massey 1999: 122).

Tourism, which was fast becoming the world's largest international industry, thus became increasingly perceived as a replacement for traditional manufacturing industries, and increasingly prioritised in urban regeneration. In 1987 a new City Council in Liverpool had adopted a more realistic attitude to local economic development, following the line that money spent on housing would never be recovered but that investing in the city as a cultural attraction would bring economic returns. In 1993 the Council appointed its first ever tourism officer. Cultural tourism was also promoted as a local development strategy by the European Union. The emphasis on the role of culture and tourism in the development of Liverpool's economy thus increased in 1994 when Merseyside was awarded Objective One status, the European Union's highest funding category reserved for its poorest regions. Nine years later Liverpool won its bid to become European Capital of Culture 2008. In 1997 the head of the Council's Department of Leisure Services declared that by 2000 Liverpool would be a 'shrine' to the Beatles (Robbie Quinn speaking at University of Liverpool in March). On 14 December 1999, Paul McCartney performed at the Cavern Club for the first time since 1963. The event was a huge coup for CCT, the local media emphasising the positive world-wide publicity it had given the city (while also providing huge publicity for McCartney's latest album) and quoted City Council officers that the city was now determined to make more of its Beatles connections. In 2001 Liverpool's Speke airport was renamed the 'John Lennon Airport' and marketed through the rather trite slogan 'Above us only sky'.

The Moptops performance event was thus staged at a time when many Liverpool policy-makers had turned to, or were being forced into considering, the economic benefits of tourism and the arts and cultural industries (Cohen 1991b, 2002), and it occupied a rather ambiguous position at the intersection of policies aimed at art, culture and the media and those aimed at tourism. There were similar developments in many other British and North American post-industrial cities. New Orleans' infamous jazz musicians and scene had attracted visiting jazz fans since the early 1940s, but the city's policy-makers associated jazz with prostitution and gambling. Only in the 1990s did they begin to promote jazz tourism. Their interest was prompted by the sudden collapse of the city's oil industry during the mid-1980s, and by the success of the New Orleans Jazz & Heritage Festival. In Memphis the former home of Elvis Presley, Gracelands, was opened to the public complete with gift shops and themed restaurants, and the city's Beale Street, site of Sun Studios where Elvis first recorded, was rebuilt and marketed to visitors as the 'birthplace of the blues' (see Connell and Gibson 2002: 246–7). In Britain Sheffield was the first City Council to develop an arts and cultural industries strategy following the rapid and dramatic decline of the city's steel industry in the early 1980s. The strategy involved long-term planning for, and launch of, the ill-fated tourist attraction The National Centre for Popular Music (Brown, Cohen and O'Connor 1998).

The growth of Beatles tourism in Liverpool can also be connected to that of a broader commercial heritage industry in Britain which emerged during the 1980s, provoking much scholarly debate (see, for example, Hewison 1987; Urry 1995; Wright 1985). This industry involved the promotion of nostalgia and the transformation of urban industrial sites into heritage centres aiming to bring the past to life in an entertaining manner, and to contribute to local economic development. It also challenged traditional notions of heritage which in Britain had tended to be rather narrowly defined, associated with the tastes and values of a conservative establishment. It thus helped to promote a British trend during the 1980s and 1990s towards broader, more pluralist notions of heritage and culture which encompassed the popular and more recent past and the everyday. This trend affected even the most notoriously conservative of institutions which had hitherto shown little interest in popular culture. Hence in 1995 Britain's National Trust announced that it would be purchasing the former Liverpool home of Sir Paul McCartney (only recently knighted) to develop it as a visitor attraction. In 2003 it did the same for John Lennon's Liverpool home. Similarly, in 1997 English Heritage, England's Historic Buildings and Monuments Commission, honoured a popular musician for the first time by erecting a plaque on the former London residence of rock guitarist Jimi Hendrix. In 2001 another plaque was unveiled on the former Liverpool home of John Lennon. In 1998 the English Tourist Board published *Rock and Pop Map of Britain: One Nation, One Groove*, a guide to places that have inspired British popular musicians. Popular music thus became legitimised as local and national heritage-based tourism.

These convergences between tourism and cultural policy increased the interlinking of music and tourism industries. CCT, for example, was a tour operator that also acted as a music agent and promoter and as a fan club. It arranged Beatles-related music events through its extensive contacts with musicians and music businesses, managed leisure facilities and organised music activities for local people and visitors. Additionally, it organized events specifically for Beatles fans, promoting meetings between fans that enabled the development of fan networks and communities. In some of its literature circulated to clients CCT promoted itself as a company acting on behalf of fans and in their interests. In fact, in 1995 CCT announced plans to launch 'the first official world-wide Beatles fan club'. As a tour operator, however, one of CCT's primary aims was to encourage Beatle fans and audiences from outside the city to visit Liverpool, thus turning fans into tourists. Meanwhile one of the main organisers of the official Liverpool Beatles fanclub explained: 'We act like a tourist office really. We are very conscious of the image of the city and we try our best to include in our literature items about what's going on in the city' (Jean Catharell, Personal Communication, 30/8/96). During Beatles Week the club organised gatherings and walking tours of the city for its members. In 1996 CCT tried unsuccessfully to obtain Objective One funding for the Mathew Street festival by defining it as a music/cultural event. At the same time, local music organisations seeking similar funding or support were pushed into justifying themselves in terms of tourism. Directors of the dance 'superclub' Cream, for example, emphasised the club's success in regularly attracting to Liverpool large numbers of club-goers from outside the city who spent money on local facilities and resources.

Convergence between tourism and music ideologies

The Moptops' performance event provoked certain discussions and debates amongst the social and occupational groups involved common to Beatles tourism, indicating similarities between discourses typical of tourism and popular music. It thus highlighted convergences between between music and tourism groups and events, institutions and policies, plus between music and tourism ideologies.

The relative merits of the Moptops and other Beatle tribute bands performing during Liverpool's Beatles Week were debated at length by

Beatles fans and by musicians, often raising the ideological distinction between representation and reality familiar to discourses of both tourism and media cultures. Distinctions are commonly drawn, for example, between tourist attractions perceived as 'contrived' or 'artificial' and those perceived as 'genuine' or 'authentic' (Stokes, 1999). The search for the real, authentic Beatles Liverpool was clearly something that motivated many fans to visit the city, and throughout Beatles Week they and other city visitors debated the accuracy of the continual stories being told about The Beatles; the authenticity of Beatles-related artefacts up for auction, or of Beatle sites; whether The Beatles should be used to represent and promote Liverpool or whether the city was 'more than The Beatles' (Cohen 1997).

Similar debates were provoked as the tribute acts were inevitably compared with the original. Among The Moptops audience were some male, British fans in their thirties, clearly irritated by Beatles tribute artists, saying they would have preferred to see 'the real thing', stressing the emphasis of earlier Liverpool conventions on video screenings of performances by the original Beatles. Other fans expressed concern about musicians who appeared to be too self-important, deluding themselves into believing that they were The Beatles, and about Beatles fans who acted as if those musicians were the real thing. Others in the audience were more concerned about how accurately The Moptops and other tribute artists had managed to recreate the original Beatles and some, along with the musicians involved, enjoyed the opportunity the bands offered to explore the tensions between musical sameness and difference. Some artists were admired for their close physical resemblance to the original Beatles, some for their skill in mimicking The Beatles' performance styles and sounds, others for the way they had somehow managed to capture the spirit of the original Beatles. As one fan put it: 'they don't try to imitate but they have the feel'.

Some of the young women who had queued backstage for The Moptops, however, were not concerned to relate the tribute bands to the original in order to judge how well they matched up, saying that attending their performances was the closest they would ever get to the real Beatles. The Moptops were thus appreciated as a vehicle that could lead to a better understanding and appreciation of the original. One small group of fans said that they liked to fantasise that The Moptops and other Beatles tribute bands actually were the original, to lose themselves in the performance and in the excitement of the event. Many obviously enjoyed the fun of the pretence thus participating with the musicians in recreating the 1960s through both mimicry and parody. With much fluttering of hands and rolling of eyes a group of George fans from Portsmouth talked about how excited they had been to meet members of the American Beatles tribute band 1964, how they had repeatedly requested 'George' to demonstrate his stage moves, and how they argued with another convention-goer who questioned the merits of the band ('He obviously wasn't a George fan'). A group of Liverpool women

from a local insurance company managed to attend an afternoon performance by a Beatles tribute band in the Cavern Club by telling their boss they
were sick. It reminded them of when they used to skip school to go to the
original Cavern, giggling when they discussed what their husbands might
think and how they might even find a new one at the performance event ('I'll
take that George in the Cavern Beatles' – giggles all round).

The Cavern Club provoked similar debates. Some visitors were unaware
that the original Cavern Club no longer existed, others criticised Liverpool
for demolishing the original saying they would have preferred to have seen
'the real thing'.[6] Some knew that the replica Cavern Club had been rebuilt
further along Mathew Street than the site of the original, in the 'wrong'
place. Other fans felt that if they couldn't see the original Cavern Club then
the reconstructions of it in both Mathew Street, and in the Beatles Story
museum in Liverpool's docklands area, were the next best thing and helped
them to feel closer to the Beatles. One of the Beatles Week coach trips
toured sites from the video for The Beatles' 'Free as a Bird' single released in
1996. At one point the coach stopped at a cluster of old, derelict and
disused warehouses, one of which had appeared in the video as the original
Cavern Club. A poster advertising the Cavern Club was displayed on its
crumbling wall, ripped and worn-looking as though it had been there for
years. The 'George fans' from Portsmouth lined up eagerly beneath the
poster pretending that they were in the famous Cavern Club queue and
urging the rest to join them or photograph them.

Tourism and music commonly raise debates about the representation and
authenticity of culture and also about its commodification. Music is a
product of what is often regarded as the most commercial and exploitative
of cultural industries (Cohen 2002), many popular music genres promoting
strong notions of authenticity, emphasising sincerity and honesty of creative
musical expression in the face of perceived commercial restrictions and
exploitation (Frith 1981; Becker 1997; Cohen 1991a; Negus 1995). Tourism
is also commonly associated with the circulation of money, thus provoking
concerns about the commercialisation of local culture through tourism and
the consequent destruction of local authenticity. Both music and tourism
industries exploit these perceived tensions between creativity and commerce,
connecting their products with notions of authenticity in order to obscure
the commercial transactions involved and increase sales (see Stratton, 1982a,
b, on the music industry).

Cavern City Tours, for example, were anxious to promote themselves as
guides to authentic local Beatles sites and experiences. The Moptops PR
leaflet stated, 'All the members of the band are sound-a-like actors/musicians
and provide authentic action and dialogue. ... Authentic Beatles sound recre-
ated live on stage!' There was, however, a fear amongst tourist entrepreneurs,
musicians and fans involved with Beatles tourism that its commercial development could destroy the authenticity of Liverpool culture. Many were
concerned that if Beatles tourism was over-commercialised the city could

become a 'tacky', 'artificial', 'plastic' 'themepark', Disneyland or Gracelands being commonly referenced as negative illustrations of what Liverpool could become. Both examples, but particularly the former, have become familiar symbols of the Americanisation of the service economy, of the commercial and inauthentic, and of fake culture with little grounding in reality. Beatles tourist entrepreneurs needed to negotiate a tension between their desire to commercially develop Beatles tourism, and their need to promote local authenticity. Similarly, many Beatles fans were critical of the idea of 'cashing in' on The Beatles, and of commercial developments that they regarded as a threat to authentic Beatles Liverpool. At the same time they wanted better organised Beatles events in terms of local facilities and resources, and for Liverpool to make more of an effort to commemorate and pay tribute to the Beatles by preserving Beatles sites and promoting the Beatles as city heritage.

These ideological distinctions between representation and the real, authenticity and commerce, have been raised by other attractions that combine both music and tourism. Fears have been expressed, for example, that tourism based around traditional and so-called 'world' music will end up destroying local music traditions, staging authenticity and inventing tradition, although others have welcomed such tourism as a means of preserving local music traditions, promoting local identity and pride and contributing to local economies. As Stokes (1999) points out, such debates typify tourism and inform people's understanding of what it is.

There were similarities and convergences between some discourses characteristic of the cultures of popular music and tourism, and also between discourses about popular music and tourism. For example, because they have been regarded as overtly commercial, both tourism and popular music have attracted a familiar mass culture critique. Music fans and tourists have been negatively stereotyped in popular and scholarly literature, being commonly perceived as unthinking, manipulated masses, or commercially exploited dupes (see Adorno 1990; Jenson 1992, on perceptions of fans). Such perceptions may help to explain the devaluing and trivialisation of tourism and popular music, and why popular music and tourism studies have been relatively marginalised within the academy (see Crick 1989 on negative perceptions of tourism within the social sciences). Members of The Moptops' audience were aware of such stereotypes and tried to distance themselves from them. Some refused to be labelled as 'fans' feeling that the way that fandom was perceived did not match their own experience (see also Cavicchi 1998); some also disliked being referred to as 'tourists': 'I'm not a tourist, I guess. I'm here because of the Beatles and if I don't have the time to visit any tourist sites then too bad, they'll have to wait until another time' (David Riesman from the USA, Personal Communication, 1996). As in other parts of Britain, local policy-makers used the term 'visitor' rather than 'tourist', to escape such negative stereotypes and to broaden out from the narrower implications of 'tourist' and encourage those travelling to the city on business, or to see family and friends, to visit city attractions.

Tourism and popular music fandom have also been debated in terms of religion (see, for example, MacCannell 1976, on tourism).). Hence Stokes (1999: 150) highlights the way that tourism has commonly been connected with pilgrimage, and music tourism has been 'framed in a religious idiom'. While many significant and insightful ideas and theories have emerged from this work, popular notions of tourism and popular music fandom as religion have again raised negative images of tourists and fans as devoted, passive and uncritical followers. In Liverpool Beatles fans were often stereotyped as obsessive, religious fanatics. A local newspaper article about the barber shop mentioned in the song 'Penny Lane' quoted Tony, the shop's owner, describing how the shop had become 'a mecca for Beatles fans from across the world': 'We get about 100 people per day visiting the shop purely through the Beatles connection. And we also get people picking away at the window frames to get a souvenir.... They believe the spirits of people are trapped in the fabric of the building which is why they go at the windows' (*Merseymart* 30 January 1997). Many Beatles fans at Beatles Week certainly described their visit to Liverpool as a 'pilgrimage'. Some also referred to the community of Beatles fans as a 'congregation', highlighting the spiritual, sacred nature of connections between Liverpool and the Beatles and telling how they had become 'born again' fans. However, as Cavicchi (1998) argues, religious metaphors and language are used by fans to stress parallels between fandom and religion rather than to suggest that fandom actually is a religion.

Conclusion and disjuncture

This chapter used a Beatles tribute band performance to explore the interrelation of popular music and tourism events, institutions and policies in British and North American post-industrial cities, to highlight convergences and parallels between the discourses that they provoked. In particular it drew attention to the growth of popular music tourism during the latter decades of the twentieth century, involving a proliferation of festivals, tours, museums, heritage quarters, and a range of different musical styles. Similar initiatives have emerged in cities not so well known for their music, or with no obvious connection to the heritage concerned. Beatles tourism, for example, exists not just in Liverpool, London and New York, but also in Prague (Connell and Gibson 2002: 285); there are now Cavern Clubs in Tokyo and Brazil. In their race to market local distinctiveness (Robins 1991), cities compete with each other to lay claim to the same music heritage title or initiative. Popular music heritage also flourishes outside the city, as illustrated by heritage theme parks and towns such as Dollywood and Branson. The music industry colludes with these developments by promoting music nostalgia to sell products, and by recycling musical trends for younger audiences.

I want to end, however, by emphasising that The Moptops performance event could also help to draw attention to disjunctures between music and tourism. There is little space here to develop this discussion adequately, but in Liverpool although popular music and tourism have converged through The Beatles, they have been more commonly regarded as entirely separate or as incompatible and conflicting. The widespread availability and appeal of commercial popular music can certainly sit rather uneasily alongside efforts of the tourism industry to promote it as local culture. Visiting Beatles fans who participated in The Moptops' performance seemed extremely knowledgeable about The Beatles and their music speaking at length about the huge significance of The Beatles to their lives and their sense of self. The Beatles were so commercially successful that their music reached and touched all sorts of people from all over the world. They became a symbol of global culture, part of everyone's heritage and the ultimate 'Music of the Millennium' (Channel 4, 1998). The Beatles' global success was celebrated with pride by Beatles tourist entrepreneurs, yet at the same time, The Beatles departure from Liverpool relatively early on in their career, challenged their efforts to promote The Beatles as Liverpool's own and to interpret and present them from a local perspective (Kaijser 2002).

Additionally, Beatles tourist entrepreneurs complained that most Liverpool residents were uninterested in, or even hostile to, their efforts to promote The Beatles as a tourist attraction. This may have been an exaggeration, but it was certainly true of many local music-makers with little involvement with Beatles tourism who attracted little recognition from policy-makers concerned with its development. Many were undoubtedly proud of The Beatles and their musical achievements, but some resented the dominance of Beatles heritage, the way it obscured or excluded other local music heritages,[7] or regarded tourism as a threat to, or even antithesis of, what music should be about. The Cavern Quarter Initiative, for example, helped to regenerate Mathew Street and its surrounding area, but the inevitable rise in rent and rates forced some small music businesses out of the area (see Zukin 1989, for account of a similar process in US cities). For those businesses, the commercial exploitation and institutionalisation of Liverpool music as heritage-based tourism sat uneasily with familiar rock ideology where music was associated with creativity, with the contemporary and everyday, with a rebellious, anti-establishment stance, with subcultures and the authenticity of 'the street', and with change rather than stasis. Many other music practitioners also argued that Liverpool should promote a vibrant, alive musical culture rather than the dead or sterile.

Music-makers in other post-industrial cities have shared similar attitudes and experiences, and the questions 'whose culture? whose heritage?' are familiar to cultural tourism in general. In New Orleans jazz musicians regularly perform for tourists, but this relationship is often regarded as exploitative (Atkinson 1996a) and 'real' New Orleans music-making has been considered as that which exists beyond the earshot of tourists and outside the city tourist areas. The city's Arts and Tourism Partnership formed during the mid-1990s to bridge the gap between arts and tourism cultures. It found, however, marked differences between the approach, motivation and even language of those involved with these cultures, which made linking the two a formidable task. Consequently, one of its initiatives involved the production of 'Flip Tips', a dictionary which translated the language of the arts into that of tourism, and vice versa, to promote understanding between those involved (Atkinson 1996b).

Finally, The Moptops' performance event problematises the familiar association of tourist culture with the visual 'gaze' and the way visual culture has been prioritised over the aural within Western society. Beatles fans participating in that event insisted that for them the most important thing was 'the music'. They described Beatles music as a powerful force, emphasising its significance at key moments in their lives, and the way that it acted as a focus, frame or backdrop to more mundane rituals and routines. They referred to music's ability to unite people, overcoming social and national divisions, to its emotional effects and its ability to move listeners and transport them from one state of being to another. For these fans the music was clearly something special plus something valued for its relevance to the ordinary and everyday. As a non-verbal, non-depictive medium with a temporal and physical dimension (De Nora 2000: 159), music was strongly perceived as penetrating personal, interior worlds and feelings, connecting with individual sense of self. It was also valued for its sociability, the way that it encourages social interaction and the construction of collective identity. Fans clearly expressed rather essentialist and romantic notions of music, evident in the way that they suggested an organic connection between The Beatles' music and Liverpool. At the same time, their discussions highlighted the significance of listening and of sonic constructions of geographical place.

Notes

1. Those involved with tourism employ numerous definitions of it, many academic researchers on tourism finding it difficult to define it in any analytically useful way (see, for example, Stokes 1999; Abram, Waldren and Macleod 1997). There is some consensus, however, that tourism can be regarded as undertaken by temporarily leisured people who voluntarily visit a place away from home for the purpose of experiencing change (Boissevain 1996, 3 quoting Valerie Smith).
2. 'Cultural tourism' covers a broad range of activities and cultural forms, but generally involves an emphasis on the tourist dimension of arts, cultural events and activities, marking a trend towards organised tourism based around more specialist interests, including heritage and nostalgia, rather than standard, packaged tours.
3. The chapter draws upon ethnographic research conducted in Liverpool between 1995 and 1997 by myself and Connie Atkinson, for a comparative project on popular music, tourism and urban regeneration in Liverpool and New Orleans. I would like to thank the Economic and Social Research Council in Britain for supporting and funding this project.
4. One hundred and thirty-four bands performed, mostly as tribute or 'cover' acts (the latter being acts who perform their own versions of songs associated with other artists).

5. They aimed, according to their manager, Mr. B. T. Blackburn, to 'recreate the magic of The Beatles' (Moptops PR leaflet).
6. At the 1996 Liverpool Beatles auction a brick claimed by the event's organisers to be from the original Cavern Club with attached plaque that explained its history was open for bids.
7. Liverpool's Capital of Culture bid of 2003 downplayed the Beatles connections in favour of emphasising the city's cultural diversity and some of its less familiar or dominant cultural attractions and heritages.

Bibliography

Abram, S., Waldren, J. , and Macleod, D.V.L. (eds) (1997) *Tourists and Tourism: Identifying with People and Places*, Oxford and New York: Berg.

Adorno,T.W. (1990) 'On Popular Music', in S. Frith and A. Goodwin (eds) *On Record*, New York: Pantheon

Atkinson, C. (1996a) 'Shakin' Your Butt for the Tourist: music's role in the identification and selling of New Orleans', in D. King and H. Taylor (eds) *Dixie Debates: perspectives on southern cultures*, London: Pluto.

Atkinson ,C. (1996b) 'Art for Tourism's Sake?', Paper presented at the Institute of Economic Studies, University of Birmingham, Nov. 7.

Becker, H. (1997) 'The Culture of a Deviant Group: the "jazz" musician', in K.Gelder and S.Thornton, *The Subcultures Reader.* London: Routledge.

Belchem, J. (2000) *Merseypride: Essays in Liverpool Exceptionalism*, Liverpool: University of Liverpool Press.

Boissevain, J. (ed) (1996) *Coping with Tourists: European Reactions to Mass Tourism*, Providence, RI: Berghahn Books.

Brown, A., Cohen, S. and O'Connor, J. (1998) *Music Policy in Sheffield, Manchester and Liverpool:A Report for Comedia*, Manchester Institute of Popular Culture, Manchester Metropolitan University, Manchester, and Institute of Popular Music, Liverpool.

Cavicchi, D. (1998) *Tramps Like Us: music and meaning among Springsteen Fans*, New York: Oxford University Press.

Cohen, S. (1991a) *Rock Culture in Liverpool: popular music in the making*, Oxford: Oxford University Press.

—— (1991b) 'Popular Music and Urban Regeneration: The Music Industries on Merseyside', *Cultural Studies*, Vol. 5 (1991), pp. 332–346.

—— (1994) 'Mapping the Sound: identity, place, and the Liverpool Sound', in M. Stokes (ed.) *Ethnicity, Identity: the musical construction of place*, Oxford: Berg, pp. 117–134

—— (1997) 'Popular Music, Tourism, and Urban Regeneration'. In S. Abram, J. Waldren and D. McLeod (eds) *Tourists and Tourism: identifying with people and places*, Oxford: Berg., pp. 71–90.

—— (1998) 'Sounding Out the City: Music and the Sensuous Production of Place'. In A. Leyshon, D. Matless and G. Revill (eds) *The Place of Music: music, space and the production of place*, New York: Guilford Press, pp. 269–290.

—— (2002) 'Paying One's Dues: The Music Business, the City and Urban Regeneration', in M. Talbot (ed.) *The Business of Music*, Liverpool: University of Liverpool Press, pp. 263–291.

—— (2003a) 'Tourism', *Continuum Encyclopedia of Popular Music of the World*, London: Continuum.

—— (2003b) 'Heritage'. *Continuum Encyclopedia of Popular Music of the World.* London: Continuum.

—— (2003c) 'Local Sound'. *Continuum Encyclopedia of Popular Music of the World.* London: Continuum.

Connell,J. and Gibson,C. (2002) *Sound Tracks: Popular Music, Identity and Place.* London: Routledge.

Crick, M. 1989. 'Representations of International Tourism in the Social Sciences: sun, sex, sights, savings, and servility'. *Annual Review of Anthropology* 18: 307–344.

De Nora, T. (2000) *Music in Everyday Life.* Cambridge: Cambridge University Press.

Frith, S. (1981) 'The Magic that can set you Free: the ideology of folk and the myth of the rock community'. In *Popular Music* 1. Cambridge: Cambridge University Press.

Hewison, R. (1987) *The Heritage Industry: Britain in a Climate of Decline.* London: Methuen.

Jensen, J. (1992) 'Fandom as Pathology: The Consequences of Characterization.' In Lisa A. Lewis (ed.) *The Adoring Audience: Fan Culture and Popular Media.* London and New York: Routledge, pp. 9–29.

Kaijser, L. (2002) 'Beatles Tourism in Liverpool'. Paper presented at the Institute of PopularMusic, University of Liverpool on 10 December.

MacCannell, Dean. (1976) *The Tourist. A New Theory of the Leisure Class.* New York: Schocken

Massey, D. (1999) 'Cities in the world'. In D. Massey, J. Allen and S. Pile, (eds) *City Worlds.* London: Routledge, pp. 99–156.

Myerscough, J. (1988) *The Economic Importance of the Arts in Britain.* London: Policy Studies Institute.

Negus, K. (1995) 'Where the Mystical Meets the Market: creativity and commerce in the production of popular music'. *The Sociological Review,* 43 (2): 316–341.

Robins, K. (1991) 'Traditions and Translation: National Culture in its Global Context', in Corner, J. and Harvey, S. (eds) *Enterprise and Heritabe: Crosscurrents of National Culture,* London: Routledge.

Stokes, M. (1999) 'Music, Travel and Tourism: An Afterword.' *The World of Music,* 41 (3): 141–56.

Stratton, J. (1982a) 'Reconciling Contradictions: the role of the artist and repertoire person in the British music industry', *Popular Music and Society,* .8 (2): 90–100.

—— (1982b) 'Between Two Worlds: art and commercialism in the record industry', *The Sociological Review,*30: 267–285.

Urry, J. (1990) *The Tourist Gaze: Leisure and Travel in Contemporary Societies.* London and Newbury Park: Sage Publications.

—— (1995) *Consuming Places.* London: Routledge.

Wright, P. (1985) *On Living in an Old Country.* London: Verso.

Zukin, S. (1989) *Loft Living: culture and capital in urban change.* Piscataway, NJ: Rutgers University Press.

7 'Troubles Tourism'

The terrorism theme park on and off screen

K. J. Donnelly

Introduction

The so-called 'Peace Dividend' in Northern Ireland in the wake of the IRA cease fire(s) has opened the door to the possibility of tourism, but under the shadow of film and television reductions of Northern Ireland. Upon my arrival in Belfast a few years ago, a taxi driver asked me whether I expected to see nothing but burning buses, barricades and petrol bombs. I replied that I never believed what I saw on television or in the cinema. Belfast has been used as a colourful backdrop to international and dramatic concerns in film and television drama. Indeed, a principal attraction of representing Belfast to the makers of films and TV drama is that it seems to offer readymade spatialized political relations. After all, a clear political-cultural division is a guarantee of drama. For tourism as much as for audiovisual drama, expectations are centrally important, and the city location *must* match them. So tourist destinations are always trying to catch up with their image, be it in holiday brochures or on electronic screens, and most of us have already been to many of these destinations thousands of times, via films and television.

Films

Films set in Northern Ireland and about 'the Troubles' constitute a genre in themselves. Belfast, in effect the capital city of Northern Ireland, is paramount to any representation of the location. British notions about the city have largely been constructed by television news, which regularly has shown wasteland and British soldiers being pelted with stones and petrol bombs. The iconographic repertoire of news reports has been limited: soldiers, armoured vehicles, armed police in bullet-proof vests, back-to-back Victorian terraces, youths throwing stones and petrol bombs. This reflects the way that stories about Northern Ireland have been limited to 'newsworthy' matters about urban unrest. Most of this iconography is highly evident in films set in Northern Ireland too.

However, the significant difference between news coverage and mainstream dramatic films about Northern Ireland is that almost all films set in Belfast have been shot elsewhere. Films are a virtual space, virtual architecture, places we can traverse mentally. Films set in Belfast, until very recently, were almost never shot there, as the cinematic version of Belfast was a *bricolage* of other cities in Britain and the Republic of Ireland. We assemble the fragments in our mind, much as we do with the actual experience of cities. As Maria Balshaw and Liam Kennedy note, 'The city is inseparable from its representations, but it is neither identical with nor reducible to them. – and so it poses complex questions about how representations traffic between physical and mental space' (2000: 3).

The problems of filming in Belfast have led to fabrications of the city, lacking a footing in 're-presentation' of the actual location. This has led to a Belfast that is more versatile for film and television drama. It is a city of the mind, where all aspects are geared towards dramatic and psychic conflict. Belfast became an imaginary landscape for 'battle' between two archetypal opposites, presented in a similar manner to the Montagues and the Capulets of *Romeo and Juliet*.

Film theorist Peter Wollen notes that: 'the great films of the twentieth-century city are often, paradoxically, studio-made films, triumphs of design rather than realistic photographic renderings of the mean streets themselves. ... The city is perceived as a kind of dream space, a delirious world of psychic projection rather than sociological delineation' (1992: 25). This description fits Belfast on screen, but for whom are these images intended? Northern Ireland-, and especially Belfast-set films and television are fantasies for British (sometimes American) audiences.[1] As Brian Neve noted, 'film representations of "the Troubles" in Northern Ireland have often seemed to perpetuate particularly British concerns' (1997: 2). Films offer touristic versions of the location, defining Northern Ireland for large audiences who have not visited in person.

Films and television dramas set in Belfast tend simply to desire the city as a colourful backdrop and a situation that allows for dramatic possibilities, divorced from explanation for Northern Ireland's political and civil antagonisms. This was noted by the British Anti-Partition of Ireland League, who issued a broadsheet at the screening of Belfast-set film *Odd Man Out* in 1947, which read: 'You have just seen, and no doubt been deeply moved by this British–Irish masterpiece ... but do you realise that the conditions prevailing in Northern Ireland have provided the background against which this terrible drama is worked out?' (quoted in Rockett *et al* 1987, 158–159). More recently, films such as *Nothing Personal* (1995) and *Resurrection Man* (1997) both reduce Loyalism to random violence, and exploit political strife for the dramatic potential of brutality rather than making any attempt to deal with sectarian ideas or their historical development in a manner that might provide the context for violence.

Fake Belfast

Northern Ireland was established as a political entity with the partition of Ireland in 1921 and has been an exception to other parts of the United Kingdom, with its own legislature until suspension in 1972 and draconian laws such as the Special Powers Act (1922), later the Emergency Provisions Act (1973). The explosion of civil violence with the Troubles in 1968 was followed by a sustained struggle between those who wanted unification with the Republic of Ireland (the Irish Republican Army) and those desiring the retention of the union with Britain (the British army [accepting the fact that the British Army has a very vocal positionon defending the legal territory], Unionists and Loyalist paramilitaries). The population is Protestant (largely Unionist or Loyalist, loyal to the British monarch and the United Kingdom) and Roman Catholic (largely Republican or Nationalist, desiring a united and independent Ireland). Protestants are the majority, comprising 51 per cent of the population of 1.6 million. Catholics number 38 per cent of the population according to figures from the Northern Ireland Office (Buckley 1997). Sectarian divisions permeate Northern Irish society, with significant paramilitary activity on both sides.

Film versions of Belfast are constructed sometimes without any reference to the actual city at all. Many films have been set in a 'Belfast of the mind', which bears only a passing resemblance to the city, yet contain exaggerated aspects that, for the purposes of film drama, are seen to be the salient features of Belfast. Consequently, aspects of Belfast that are not a part of the dramatic scenario, and that do not register on a global level, are elided, ignored and deemed not worthy of presentation. For example, the city's riches and opulence that were born of thriving heavy industries until the 1970s, are only rarely recognized by films. These aspects would detract fatally from film and TV representations of Belfast as a desolate urban battleground.

Films shot in Belfast

Director of the Northern Ireland Film Council, Geraldine Wilkins, noted that 'no indigenous feature was made in Northern Ireland between the 1930s and the 1980s' (Wilkins 1994). A small number of films made by British interests have had significant sections shot in Belfast. The most noteworthy are *Odd Man Out* (1946), directed by Carol Reed and starring James Mason; *Jacqueline* (1956), partially scripted by Catherine Cookson and starring John Gregson; *Maeve* (1981), directed by Pat Murphy; *Divorcing Jack* (1998), adapted by Colin Bateman from his comic novel and starring David Thewlis; *Titanic Town* (1998), starring Julie Walters. In addition to these, local interests have been involved in shooting Channel Four television drama *Acceptable Levels* (1984) and BBC serial *Eureka Street* (1999 BBC); the last of these was adapted from a novel by Robert McLiam Wilson and made a feature of its Belfast locations.[2] Occasionally, films have been shot outside

Belfast. One such film was *December Bride* (1990), directed by Thaddeus O'Sullivan, which was a costume drama set around 1900 and one of the only films not directly dealing with 'the Troubles'. It uses Dublin locations for what was meant to be Belfast despite being shot on location at Strangford Lough, a short distance from Belfast, and illustrates the entrenched tradition of shooting films elsewhere as 'Belfast'.

There are many films that have been set in Northern Ireland but shot elsewhere. These include *Ascendancy* (1982), *Angel* (1982, directed by Neil Jordan), *Cal* (1984), *Hidden Agenda* (1990, directed by Ken Loach), *In the Name of the Father* (1993, directed by Jim Sheridan), *The Crying Game* (1993, directed by Neil Jordan), *Nothing Personal* (1995, directed by Thaddeus O'Sullivan), *Some Mother's Son* (1996, directed by Terry George), *The Devil's Own* (1997, directed by Alan Pakula), *Resurrection Man* (1998) and *The Boxer* (1998, directed by Jim Sheridan. This is not an exhaustive list. It demonstrates, however, how far The Troubles in Northern Ireland have proved an attraction for film-makers and a staple for audiences outside Northern Ireland. (Donnelly, 2000) Typical of a British production, *Resurrection Man* was shot in Manchester, Liverpool and Warrington.

The 'peace dividend' has allowed filming in Belfast. Indeed, five features were shot in Northern Ireland in 1997 (Adams 2000). Moreover, the 'peace dividend' has opened the possibility of tourism in Northern Ireland, beyond the basic idea of visitors fishing in Fermanagh in the west of the province, to tourism in Belfast itself. Tourist numbers grew significantly after the IRA ceasefire (from 1994–96, and since 1997).[3] Tourism has always been problematic in Northern Ireland. Visitors will find a dichotomy of mementoes on sale. Either there is expensive lace, and whiskey produced by Bushmills, or there are cheap tourist items that involve leprechauns and shamrocks. The latter cater for an undifferentiated tourist version of 'Ireland' and in the vast majority of cases are made in China. In fact, the most specific aspect of Northern Ireland is The Troubles itself; its 'USP' (unique selling point), borne out by the staggering amount of local books produced that deal with Northern Ireland's heritage of sectarianism and civil conflict. In addition to this, we should note the many touristic versions of Northern Ireland, and Belfast in particular, that overseas audiences have 'visited' on screen. Films and television define Northern Ireland for large audiences who never visit.

LOCATIONS

Perhaps the most prominent film shot largely in Belfast was *Odd Man Out* (1947). Despite the foregrounded location shooting, the film tried to channel meaning into certain directions and away from others. *Odd Man Out*'s opening rolling title card is a heavy-handed attempt at delimiting meaning. It reads:

> This story is told against a background of political unrest in a city of Northern Ireland. It is not concerned with the struggle between the law and an illegal organisation, but only with the conflict in the heads of the people when they become unexpectedly involved.

This caveat aims to excuse the film from providing precise depictions, as well as attempting to remove the emphasis from the IRA and civil conflict. Thus, *Odd Man Out* tries to ignore the historical precedents and sedimentations of the film's location, characters and their motivations. And yet, simultaneously it works against this. Images of the location contain highly-charged meaning – for certain audiences – by their merest depiction. Opening shots of the film are of two significant Belfast locations: Cave Hill and the Harland and Wolff shipyard. These not only explicitly indicate Belfast as the film's setting, but they also delineate the sectarian divide of Northern Ireland through these two charged images: Cave Hill, the symbolic site of the 1798 rebellion, and the shipyard that was a symbol of Ulster Protestant power and almost exclusive employment. The film's story centred upon the Albert Clock and concluded at the Customs House at the river. The bomb shelter part of the film was shot in the Falls area of the city, at the back of Springfield Road near Forfar Street. Among locations that appear in the film, there is the Crown liquor saloon on Great Victoria Street, which now is a heritage site adorning postcards. Both for tourists and screen tourists, cities are 'metonymised'. They are reduced to key sites: a city's 'essential' or 'distinctive' elements. For example, the Ordnance Survey of Northern Ireland map of Belfast (copyrighted 1994) sports on its cover photographs of Queens University, Harland and Wolff shipyard, Belfast City Hall, Belfast Castle and the Castle Court shopping mall on Royal Avenue.

Films shot in Belfast have stuck to the same few indicative locations: Harland and Wolff, Cave Hill, Botanic Gardens, Ulster Museum, the Crown Bar and the Waterfront Concert Hall. The world famous Harland and Wolff shipyard, with its two enormous gantries, is easily visible from large areas of the city. Apart from the opening shot of *Odd Man Out*, McNeil (John Gregson) works in the shipyard in *Jacqueline* (1956). Cave Hill overlooks Belfast from the west. This has remained in Republican lore as the site of the conspirators for the 1798 rebellion, the 'United Irishmen'. The most obvious landmark is the square pediment known as 'McArt's Fort'. This is shown in the opening shot of *Odd Man Out* and also appears in the distance in *Divorcing Jack* and a few times in *Titanic Town* and *Eureka Street*. Belfast City Hall is the central focus of the city's central business district. In most cities the city hall provides an instant indication of the city and municipality to those who are familiar with the locality. It appears prominently in the opening title sequence of *Divorcing Jack* and includes a shot of the statue of Queen Victoria that fronts the city hall, pulling back to include the whole of city hall in the shot and then follows the film's central character, Dan Starkey played by David Thewlis, as he walks away down Donegal Place with City Hall frontage in the background. The Botanic Gardens appear only in *Divorcing Jack*, although on both occasions avoiding the distinctive Palm House and Ulster Museum.

The museum is a building that seems crudely to echo the province's schizophrenia in its mixture of Victorian brick and brutalist concrete. Its historical exhibition about Northern Ireland only runs from Plantation in the seventeenth century to the establishment of the 'statelet' of Northern Ireland within the United Kingdom in 1921. So, a gapped 'official' history like this makes people into tourists in their own home, able to see the unseen/unwritten about their city. The Crown Bar is a Victorian hostelry with characteristic cubicles, on Great Victoria Street. It appears in *Odd Man Out,* where injured IRA man Johnny is left alone in a cubicle and hallucinates bubbles containing talking faces. It also appears briefly in *Divorcing Jack*, where Starkey and American journalist 'Charles Parker' drink whiskey. The inclusion of this journalist functions at least partially to allow tourism within the film, allowing a view of Belfast's 'sights'. The Waterfront Concert Hall, which opened in 1997, is most famous for being the venue where U2 did the 'Yes' concert during which politicians David Trimble and John Hume appeared on stage and shook hands. The foyer is where protagonist Starkey attends politician Brinn's press conference in *Divorcing Jack*. Oddly, that film has no establishing shot of that building (or for its sequence shot in the Crown Bar). The Waterfront Hall also appeared prominently in *Eureka Street*. It was where Chuckie gave his successful pitch to the Northern Ireland Development Corporation, and allowed an impressive vista of Belfast out of the large glass window.

BBC serial *Eureka Street* included a startling 360-degree pan shot from a tall building in Belfast. This took in many of the city's landmarks (most obviously McArt's Fort on Cave Hill and the City Hospital) and gave a striking sense of unity to the city and its representation in the programme, underlined by protagonist Jake's description of Belfast as 'my city'. This was the final shot of the drama, and was a breathtaking audiovisual flourish of crane shot and panorama, harmonising the divided city under the central character's poetic voice-over eulogy to Belfast. Camera crews are very obvious 'tourists', and are afraid of a negative reaction from locals. They do not want to be seen in Belfast, and hence there are few establishing shots in these films or dramas. The difficulties of shooting in Belfast were illustrated by the problems experienced in filming *Divorcing Jack*'s scene at the Botanic Gardens. This was the same day as Tony Blair's historic handshake with Gerry Adams, and shooting was constantly interrupted by 'a fleet of military helicopters' overhead. (Adams op. cit.). While *Divorcing Jack* shows many Belfast 'sights', it does not show key sectarian areas, such as the Falls Road, Shankill Road, Newtownards Road or Andersonstown. Although it was shot in Belfast, the production was forced to stick to fairly 'neutral' areas, in fact those that tourists might see. But the overall effect of the lack of establishing shots in these films is that locals recognize the location, visitors might and 'outsiders' very likely will not.

Significantly, there are a number of important landmarks of the city of Belfast that are generally left unrepresented in cinematic views of the city. One is Milltown Cemetery, the place where many IRA volunteers are buried and the scene of Michael Stone's attack on mourners in March of 1988 (Dillon 1993). This has to be seen as a key Republican location in Belfast, set in Republican west Belfast and between the Falls Road and Andersonstown, both Catholic strongholds. Another unrepresented location is Queen's University of Belfast. Once, this would have been a symbol of Unionism, but in more recent years, it could have been used to represent a neutrality in the sectarian warfare, as the Botanic area of the city tends to be seen as a student area without specific ties. Some of the 'essential elements' used to represent or concretise Belfast in tourist guidebooks are not the same as the urban iconography used in the films and on television.

Landmarks

Essential landmarks are 'sights' and a tourist object. They *must* be seen. Distinctive locations are a city's important 'sites', and are converted easily into film and television iconography. These might be more vague but they define a specific location, especially an urban one. In the case of Belfast, these would have to be the city's sectarian murals and its 'peace lines' '(peace walls)'. The fact that murals are only marginal to films that fabricate Belfast suggests that these films are not interested in the key signifiers of identity for Belfast people; they are simply interested in using the conflict for dramatic reasons for audiences overseas. Bill Rolston's books, which compile photographs of these murals, are displayed prominently in Belfast bookshops (Rolston 1992, 1995). 'Peace lines' are a defining characteristic of the Belfast landscape, dividing Catholic from Protestant and serving to stabilize the urban through an accepted apartheid. There are twenty-six of them in the city. However, 'peace lines' are shown only in films that have been shot elsewhere. A large metal gateway appeared in the distance in *The Devil's Own* (1997). *Nothing Personal* has John Lynch's character cross a peace line, suggesting the close proximity of Catholic and Protestant areas, like around Donegal Pass, or around Springfield Road. *Cal* shows the 'ring of steel' of checkpoints skirting Belfast city centre, where John Lynch's character (a different one) has to be checked by soldiers when passing through. However, films such as *Nothing Personal* and *Resurrection Man* show *no* Belfast landmarks, not even as insert shots. This underlines their status as landscapes of the mind, and ones produced for the consumption of people outside of Northern Ireland, what we might call 'screen tourists'.

'Armchair' or 'screen' tourism is evident in all the holiday programmes on television, which effectively are mini-holidays (in fast-forward) by proxy. This is a more eco-friendly form of tourism, where places are not worn away by a constant stream of tourists, and local cultures are not rebuilt around tourist visions. All screen visits to other locations might be construed as a form of 'tourism'. This is the point where tourism meets neo-colonialism. There is a great tradition of travelogue films, something that was very common in the early years of cinema, after which it became largely submerged into mainstream dramatic films. Television currently betrays a touristic obsession with location, as exemplified by *Heartbeat, Peak Practice* and *Ballykissangel* (Nelson 1997: 46). In the wake of these programmes, shops and support industries have grown up where these are shot to sell the location of the show to visitors.

Troubles tours

Belfast has shops that match the city's sectarian divide and sell the cultural form of civil conflict. The Republican shop, run by Sinn Fein, is on the Falls Road at the heart of the city's Catholic community in west Belfast. It is a fairly upmarket establishment, selling books of Bobby Sands' poetry, political literature and T-shirts, among other things. One cannot help but imagine that it is aimed at least partially at American tourists. Its equivalent on the Loyalist side is on the Newtownards Road in Protestant east Belfast. In contradistinction, this is a cheaper bazaar, selling paramilitary mugs and baby bibs, along with theme candies shaped as red hands or orange feet.

Belfast's sectarian divisions and peace walls make it a difficult city for the tourist to traverse on foot. Once, while walking through Protestant east Belfast, I wandered into the small and beleaguered Catholic enclave of Short Strand. This was a potentially dangerous thing to do, especially as tensions were fairly high at the time. It is difficult to approach the Catholic Falls area of the city from the Protestant-dominated area around the City Hospital on Donegal Road. The pedestrian has actually to walk through the Royal Victoria Hospital to reach the Falls Road from this direction. However, Sinn Fein city councillor Tom Hartley occasionally leads walks of west Belfast, explaining historical and cultural aspects. Along similar lines, Failte Feirste Thiar, a west Belfast community group dealing with tourism, made available a multi-lingual map detailing some of the historical sites of violence in the city (Tourists' Troubles Map 2001). According to the *Guardian,* 'The map's organisers were keen to emphasise that they were not trying to turn the area into a terrorist theme park' (Gentleman 1999). I went on a minibus tour that started at the YMCA and was led by the friendly and informative driver 'Rodney'. It was an intriguing tour of the key sites of the Troubles, both past and present. Only a few people were on the bus and it seemed less official than other bus tours.[4] We were also given a sense of some drama during the tour, when Rodney told us that we would have to 'watch out for stone throwing' – and that we might be the target! An American newspaper ran a review of the tour. Although there was some confusion about the differences between the Republican and Loyalist sides and the (seeming) mistake of calling the Sinn Fein office the 'IRA Headquarters' (Lansky 1999). This tour certainly carved out a very particular experience of Belfast . One problem for the tours is this: which 'reality' should they present? Loyalist? Republican? Television, and hardly to a lesser degree films, offer a British view (sometimes an American view, or view from the Republic of Ireland) that (oddly) sees itself as a 'neutral' ground – from *outside* the field of play. This is the target audience for 'Troubles Tourism' as much as it is for film and television representations. However, the door of tourism is definitely open to global capital. On the Shankill Road, I saw and photographed a half-painted Loyalist paramilitary mural on the side of a Kentucky Fried Chicken outlet. The paramilitary retention of the red colour scheme, and the franchise's willingness to allow this is intriguing. I am unable to resolve whether this is a situation of local corrupting global capital, or global capital finding an accommodation with sectarian violence.

Conclusion: an-Other world

Belfast film illustrates cinema's more general situation, where it is more about the construction and packaging of another world (or rather an 'Other' world) rather than a simple representation of this one. Film depictions of Belfast remind me of Giorgio de Chirico's paintings, Anton Furst's Gotham City sets in Tim Burton's *Batman* (1990), and Jean-Luc Godard's futuristic city constructed from collages of brand new Parisian buildings in *Alphaville* (1965). Peter Wollen may well be right about the most enduring city films being fabrications rather than travelogues, but I am still concerned as to who the target audience might be for such audiovisual representations, or rather packagings, of Belfast and the Troubles.

The answer has to be for people from elsewhere, for tourists. The 'Troubles Tourism' industry is being determined from outside, by expectations established and succoured by film and television. According to John Wilson Foster, 'script-writers and playwrights ... have commodified the city of Belfast, indeed all of Northern Ireland, by reducing it to one relentless image: that of a Catholic, occasionally Protestant working class eternally turbulent and besieged. As a result, many in Ulster are estranged from their own image; they have no mirror in which to see themselves' (Foster 1992: 169).

Northern Ireland on film appears as British violence fantasy, where manipulation of the periphery holds the cultural centre, and asserts the hegemonic (perhaps even colonial) power relation of the regime of representation. The 'Peace Dividend' has started to move the sectarian conflict in Northern Ireland even further onto the cultural stage than it already had been. Hence tours are culturally rich, but they increasingly will find it difficult to avoid duplicating film and television spectacles. Consequently, they will drift towards film and television presentations of Northern Ireland as irrational and beyond explanation.

Coda: Belfast diary, July 2000

In itself, this paper constitutes a form of tourism. So it seems fitting to finish it with a brief excerpt from my diary while in Belfast. It illustrates a fraught week for the peace process in the wake of the Parades Commission's refusal to allow an Orange march along the Garvaghy Road in Portadown. It also illustrates how close scholarly research and uncomfortable cultural tourism can be.

Monday 3rd

The Orange Order's call for protest yesterday at Drumcree meant that expectation about trouble on the streets of Belfast is high. Visit the 'Political Collection' held at the Linen Hall Library. Librarian Yvonne Murphy and her small team have done an excellent job collating massive amounts of material concerned with the Troubles. Among their impressive holdings are items such as miniature issues of Republican News for smuggling into prisons, and a medal for the siege at Drumcree with 'siege' spelled the wrong way. Access is limited to researchers who make arrangements with them. It could prove a focus for positive developments in Northern Ireland, embodying the very rich sectarian cultures and offering a massive potential for education. Cultural and educational possibilities are already coming to fruition as part of the 'peace dividend'. Yvonne tells me that the Brighton bomber's PhD thesis is soon to be published, and on the way out I pass a few words with a former Loyalist prisoner who is embarking on a degree course.

Tuesday 4th

Visit the Republican shop on the Falls Road and the Loyalist shop on Newtownards Road. The Republican shop is almost like any upmarket tourist shop, while the Loyalist one is like a downmarket bazaar, where I think I see Johnny Adair, recently-released leader of the UFF paramilitary group, who has been on television and in the newspapers a lot lately. Particularly happy to buy a copy of the Loyalist dance mix CD which is not available at record shops, although I can't find any of the orange foot-shaped lollies emblazoned with 'Drumcree'. The Loyalist shop is full of people milling around and some Scots who assure the counter assistant that they will be at Drumcree tonight. I photograph many impressive murals during my walk. Loyalist murals almost always include gunmen, who invariably are sporting Adidas tracksuits. Is this some form of sponsorship? On the way home, walk past one end of the Loyalist Donegall Pass, where mothers and young children have formed a peaceful road block. Later, on local television this scene is replaced by burning cars. I watch on television what I could walk to in two minutes.

Wednesday 5th

Researching at the Political Collection again. In late afternoon, I walk through people spilling out from a graduation; ceremony at Queen's University. Next week, it's the same at my university. Look out of hotel window and see plumes of smoke slowly curling upwards. There is no wind and it's obviously a lovely warm evening for making barricades of burning cars.

Thursday 6th

Visit the Public Records Office (PRONI). Go on a bus tour of the city in the afternoon. This provides much information about architecture and history, particularly history of the troubles and how it has affected the environs of Belfast. British television news says that all is quiet in Northern Ireland. Have trouble hearing the report for police sirens as armoured land rovers speed past.

Friday 7th

Visit Belfast city library and the Linen Hall library. It seems to take hours just to peruse the massive numbers of books that are published about Northern Ireland, a large proportion of which are printed by local publishing houses. Later visit the Ulster Museum. The historical section only runs from immediately before the introduction of Protestant settlers in the 16th century to the formation of Northern Ireland in 1921. The exhibition is very incoherent, and interesting precisely for this reason.

Saturday 8th

There are far less people shopping in the city centre than you'd expect. Go on a Citybus tour which takes in the most 'troubled' areas. The driver is very informative, detailing the atrocities of the Shankill Butchers and showing us the 'Murder Mile' in north Belfast. This had a little noted mention in Elvis Costello's number one record, 'Oliver's Army'. At one point in the tour, the driver allows me to get out of the bus and photograph a recent Loyalist mural that celebrates the killing of Catholic civilians. I do it very quickly, in fact I haven't moved so fast in years. I also get a great photograph on the nearby Shankill Road. Here, a bright red mural with black gunmen figures remains half-painted on the side of the red and white colours of a Kentucky Fried Chicken outlet.

Sunday 9th

Belfast's shops are all closed on Sundays, unlike most of the UK. Go on a (heavily curtailed) bus tour to the Giant's Causeway. The roads are empty and everyone is wondering what happened at the Drumcree march today. British television says that things are now calm and Drumcree is under control. On the streets, almost all eating places appear to be shut and the regular police patrols show that trouble is on the way.

Monday 10th

The Orange order has called for a 'strike' for four hours, from four o'clock till eight. I'm warned to leave early to avoid the unofficial road blocks. Notice that from about 2 o'clock the city looks like a frantic rush hour with all the shops closing and everyone leaving for home. Decide to walk, carrying heavy baggage, against the flow of pedestrians and crawling cars. The people look like a flood of refugees. My gait is awkward, and I stagger like James Mason in *Odd Man Out*. Like him, I end up at the Customs House. It is just after 3 o'clock and the nearby bus station has 'suspended all services'. I walk the rest of the way into the docks. Leaning over the side of the Liverpool ferry, I watch the antics of black guillemots in the water and see black smoke rising over east Belfast, and wonder what might happen were I to be stopped and found with both Loyalist and Republican ephemera.

Notes

1 It is interesting to note that between 1995 and 1997, despite overall numbers of visitors to Northern Ireland from Britain being 585,000 and from the USA being 85,000, those visiting on holiday from each country were the same number at 37,000 (McArt and Campbell 1999).
2 Other films that have partly been shot in Belfast include *Born for Hell* (1976), a Canadian film in French that was also shot in Germany, and *With Or Without You* (1999), a British-made romance starring Dervla Kirwan and Christopher Eccleston and directed by Michael Winterbottom.
3 From 1,262,000 visits in 1993 to 1,557,000 visits in 1995 (Northern Ireland Annual Abstract of Statistics 1998: 84).
4 Some black taxis also provide a tour of west Belfast.

Bibliography

Adams, M. (2000) *Movie Locations: A Guide to Great Britain and Ireland*, London: Boxtree.
Balshaw, M. and Kennedy, L. (eds) (2000) *Urban Space and Representation*, London: Pluto.

Buckley, R. (ed.) (1997) *The Irish Question: Division and Reconciliation*, Cheltenham: Understanding Global Issues.

Dillon, M. (1993) *Stone Cold: The True Story of Michael Stone and the Milltown Massacre*, London: Arrow.

Donnelly, K. J. (2000) 'The Policing of Cinema: Troubled Film Exhibition in Northern Ireland', *Historical Journal of Film, Radio andTelevision*, 20 (3).

Foster, J. (1992) 'Culture and Colonization: A Northern Perspective'. In Michael Kennedy (ed.) *Irish Literature and Culture*, Gerard's Cross, Bucks.: Colin Smythe.

Gentleman, A. (1999) 'Tourist Guide to Belfast Troubles', *Guardian,* Monday 2 August <http://www.guardian.co.uk/Northern_Ireland/Story/0,2763,204844,00.html> (accessed 9 October 1999).

Lansky, D. (1999) 'Political Tour Covers Belfast's Troubles', *Charlotteville Vagabond,* Sunday 30 May <http:// charlotte.com/justgo/travel/0530belfast.htm> (accessed 9 October 2001).

McArt, P. and Campbell, D. (1999*) Irish Almanac and Yearbook of Facts*, Burt, Donegal: Artcam.

Nelson, R. (1997) *TV Drama in Transition: Forms, Values and Cultural Change*, Basingstoke: Macmillan.

Neve, B. (1997) 'Cinema, the Ceasefire, and "the Troubles" ', *Irish Studies Review*, 20, Autumn. *Northern Ireland Annual Abstract of Statistics* (1998) Belfast: Northern Ireland Statistics and Research Agency.

Rockett, R., Gibbons, L. and Hill, J. (1987) *Cinema and Ireland*, Kent: Croom Helm.

Rolston, B. (1992*) Drawing Support: Murals in the North of Ireland*, Belfast: Beyond the Pale.

——(1995) *Drawing Support 2: Murals of War and Peace*, Belfast: Beyond the Pale.

Tourists' Troubles Map (1999) BBC News, BBC Online Network, Friday 30 July 1999 <http://news.bbc.co.uk/ hi/english/newsid_407000/407243.stm> (accessed 9 October 2001).

Wilkins, G. (1994) 'Film Production in Northern Ireland', in J. Hill, M. McLoone and P. Hainsworth (eds) *Border Crossing: Film in Ireland,Britain and Europe.* Belfast: Institute of Irish Studies/BFI.

Wollen, P. (1992) 'Delirious Projections', *Sight and Sound*, 2 (4).

8 Mediating William Wallace

Audio-visual technologies in tourism

Tim Edensor

People are tourists most of the time whether they are literally mobile or only experience simulated mobility through the incredible fluidity of multiple signs and electronic images

(John Urry 1995: 148).

Certain dystopian commentators regard contemporary Western, 'postmodern' culture as increasingly centred around visual experience in which subjects become lost in a blizzard of signs. Accordingly, this welter of popular signifiers, detached from that which they signify, are simulations of an ungraspable, even chimerical, reality (Cubitt 2001). The profusion of commodity signs, adverts, images of 'exotic otherness', filmic and televisual imagery, flow across space, detaching any sense of meaning and belonging to place, flattening localities in an homogenising process which turns all situated cultures into spectacles and simulacra. This sign culture would seem to be nowhere more prevalent than in the tourist industry, depending as it does upon the marketing of distinctive cultural features to attract tourists, often allied with strategies devised to attract inward investment, shoppers and key middle-class professionals through revamping and rebranding places. John Urry has identified tourism as primarily concerned with visual consumption, suggesting that the gaze dominates tourist activity, although it takes different forms. A prevalent mode of tourist gazing is the 'mediatised gaze', shaped by rebranded places and re-presentations of heritage which draw upon films and television programmes (2002: 151). Clearly, the simulation thesis appositely identifies expanding commodification in a globalising world and the ways in which this stimulates the rebranding and manufacturing of place-images.

A vast engagement with the commodification of images and signs then, is prevalent in the international tourist industry. Integral to the mediatised production of place is the multitude of dramatised productions at tourist and heritage attractions. Besides the small playlets, dramaturgical re-enactments and performances staged for tourists' 'edutainment' (see Edensor 2001), such sites also increasingly feature audio-visual presentations, *son et lumiere* shows and animatronic characters which add to those already existing visual associations with place. With a displacement of 'authentic' artefacts on display, the contemporary tourist increasingly enjoys immersion in a more figural, affective and mediated experience. It may appear as if this tendency highlights how the attractions that increasingly emphasise the visual and the dramatic, are akin to those qualities associated with movies, as opposed to the didactic presentation of historical evidence and authoritative accounts. However, it is vital to beware of removing the agency of consumers, dematerialising space and disembodying tourists. Accordingly, I will explore whether the movie screen is really analogous to the diverse audiovisual presentations offered at tourist attractions by investigating the ways in which tourists variously mediate place and history, and speculatively assessing the effects such installations may produce.

My aim in this paper is to explore the recent re-presentation of a well-known Scottish tourist attraction – the Wallace Monument – which has overhauled its delivery of information and entertainment in compliance with this imperative towards audiovisual techniques. Pertinent to this study, this site has also had its profile raised considerably by its association with the Hollywood blockbuster, *Braveheart*, further adding to the mediatisation of the Monument. Both film and monument centre upon William Wallace, a prototypical 'freedom fighter' for Scottish independence in the medieval era, who has long been one of the greatest mythic heroes of Scotland despite the dearth of historical knowledge about his life and times. The struggle against English occupation initiated under the leadership of Wallace, and resulting ultimately in the achievement of Independence in 1314 following the Battle of Bannockburn, in which a Scottish army led by Robert Bruce defeated a larger English force, retains a metaphorical power for many Scots despite the historical distance from that era. And although Wallace was caught and brutally executed by the English, he remains an exemplary figure of courage and patriotism for many Scots. Inaugurated in 1861, the Wallace Monument testifies to an upsurge of nineteenth-century romantic nationalist sentiment, and has since then been a prominent Stirling landmark and a site of pilgrimage for nationalists and other tourists.

Mediating landscape

The Wallace Monument specifically commemorates the Battle of Stirling Bridge which took place in 1297, close to the site of the monument, where a Scottish force commanded by Wallace defeated an English army, and evidently mediates the meaning of Wallace in its built form. Most obviously, it uses the semiotic conventions of memorial architecture to materialise and mediate the importance of Wallace as an icon. Built in the Scottish baronial style – signifying grandeur and import – atop the wooded Abbey Craig hill, it transmits an idea of heroism for miles around. Together with Bannockburn Heritage site and Stirling Castle, the Wallace Monument forms a third point of the 'Stirling Triangle', the historical features which mark Stirling out as significant for many Scots and tourists alike. Here then, the landscape has been textually inscribed and reinscribed as significant, and is already a mediated space, especially evident when a Stirling panorama is beheld from a number of vantage points. This mediation via landscape should not be neglected for space is devisedly etched and encoded with points of significance to convey a range of ideological meanings which are intended to be read. Why else would monuments get built? Inevitably, such space-texts are open to interpretation and hegemonic and preferred encodings may well be challenged or subside over time (Warner 1993). Nevertheless, while each landscape is unique, there are conventions which bear upon the ways in which landscape is written, and memorial landscapes require for their recognition a familiarity with other landscapes and places. As Barnes and Duncan have maintained, space is never apprehended as a blank sheet devoid of context: 'various texts and discursive practices based on previous texts are deeply inscribed in ... landscapes and institutions' (1992: 8).

The Wallace Monument lies within a network of tourist attractions, representations and narratives, and other heterogeneous associations. In one sense, it is part of a series of local attractions which also proffer mediatised versions of heritage through audiovisual presentations. The nearby Rob Roy Centre in Callander is wholly shaped by its audiovisual element. Located in an old church, its main attractions are a cinematic tour of the Trossachs and installed in a Highland diorama, a full-scale, animatronic model of Rob with a fellow clansman discussing his feud with the Duke of Montrose. At Bannockburn Heritage Centre, a theatre is devoted to a video display of the Battle of Bannockburn, and the Loch Lomond Centre has two presentations, one a fish's eye view of the loch's ecology and its legends, the other focusing on the story behind the famous song, *The Bonnie, Bonnie Banks of Loch Lomond*. Thus the tourist realm in which the Wallace Monument is situated is already saturated with images, ideas and narratives in addition to those from film, television and tourist marketing publications.

The *Braveheart* effect

Allied to these inter-media effects, as mentioned the Wallace Monument has been given an enormous boost in popularity by the film *Braveheart*, which purports to tell the story of Wallace. I will briefly focus upon this before exploring the effects of the monument's other audiovisual fixtures. Mel Gibson directed and starred in the film which takes a somewhat loose approach to historical veracity, foregrounding some of the major battles Wallace was engaged in as well as furnishing viewers with some spurious romantic details. Attendance at the monument more than doubled in the year following the film and the site has developed a higher profile for foreign and domestic tourists, a phenomenon which was also accompanied by the revamping of the monument.

Braveheart is undoubtedly part of what Appadurai refers to as global 'mediascapes', being specifically part of the production of images which transmit notions of difference and stimulate the 'desire for acquisition and movement' (Appadurai 1990) – desires exemplified by tourism. Hollywood films are devised to appeal to 'fantasies, desires and aspirations that are not simply of local and national interests', but are integral to the film cultures of most parts of the world (Higson 1995: 8). Hollywood casts its net wide, often alighting on mythical heroes and epics from other cultural locations (for instance, *The Three Musketeers, El Cid, Robin Hood*), thereby disembedding the telling of these stories from localities and encoding them with its own stylised themes and versions of romance, masculinity and femininity, and individual courage and integrity. The circulation of these images and narratives of 'otherness' signify familiar notions of global difference. According to German film-maker Edgar Reitz, Hollywood has 'taken narrative possession of our past' (Morley and Robins 1995: 93). With broader regard to the Hollywood portrayal of Scottishness, certain distinct tropes can be identified. For instance, the 'kailyard' tradition – small town or rural whimsy – is well represented in *Brigadoon* and *Local Hero*, and also features in television series *Take the High Road* and *Hamish McBeth*. There are also well-worn representations of Glasgow 'hard men' and romantic, kilted warriors (see McArthur C (ed.) 1982 for a critical account of these representations). But besides these markers of distinctiveness, Hollywood also tends to reconfigure stories within well-worn representative conventions of romance, individualism and heroism. In this sense, *Braveheart* was referred to by some commentators as a Scottish 'Western'.

Braveheart was undoubtedly a somewhat formulaic product intended for a wide global audience, and yet to assume that the film was merely a blank, stereotypical cypher for Scottishness which dazed movie-goers into passivity could not be further from the truth, at least as far as its reception in Scotland was concerned. For *Braveheart* played upon renewed desires for Scottish autonomy and independence, causing an enormous stir in Scotland where it was eagerly grasped by politicians and commentators of all political hues for its metaphorical message. Socialists claimed that Wallace was a 'man of the people', nationalists drew strong parallels between the medieval English colonisation of Scotland and current Scottish subservience to Westminster, whereas Tories were willing to consign him to insignificance. There was also great debate about the contemporary relevance and exemplary virtues of Wallace within the popular press. In a sense then, *Braveheart* was repatriated and recontextualised by the Scotland from which it had emerged (see Edensor (1998) for a detailed account of these responses). Crucially, the Wallace myth predates *Braveheart* and continues to possess a potency which means that the hollowed-out Hollywood Wallace must always stand for far more than the characterisation and tropes addressed in the film. I consider the film to be another node in the constellation of images, narratives, spaces, artefacts and rituals that surround Wallace and

provide an intertextual storehouse that is a resource for countless appropriations and interpretations of the figure's significance (see Edensor 2002).
Adding yet further layers to the numerous historical uses of this iconic character, this highlights that mythic symbols are rarely overdetermined. Indeed,
the potency of the Wallace myth depends on this flexibility, for as Samuel
and Thompson have asserted, such ideologically 'chameleon' forms (1990: 3)
can be used to transmit contrasting messages and identities. Serving as a
'condensation symbol' (Cohen 1985: 102), Wallace has been (re)appropri-
ated by a wide range of contesting groups to provide antecedence and
continuity to a diverse range of identities and political objectives.

As mentioned, the popularity of the movie meant that visitor numbers to
the Wallace Monument have greatly increased since the film's release.
Indeed, marketing strategies focused on the centrality of Stirling to the
Wallace myth, advertising the fact that 'Stirling is Braveheart country' to
maximise the potential of this association in an economy of tourist signs
and further mediating the monument and Wallace.

To summarise what has been said so far, the Wallace Monument is
already a complex constellation of different media forms, enmeshed within
associations drawn from the landscape, film, literature, architecture and
sculpture. Links have also been forged through exhibitions about Wallace,
paintings, literature, souvenirs and dramas (Edensor 2002). Now with the
recent development of several audiovisual displays and interactive technologies located within the monument, the building and Wallace have been
subject to further layers of mediation. This may well exemplify the ways in
which an 'array of tourist professionals ... attempt to reproduce ever-new
objects of the tourist gaze' (Urry 1995: 133). But many of the tourists who
visit the monument will come with prior sentiments and understandings
about the hero drawn from the wealth of mediated forms which predate the
new additions.

Wallace as Mel

The Wallace Monument is a complex attraction containing a variety of
different forms of information, modes of engagement and kinds of space.
There is a veritable mix of audiovisual provision in the Wallace Monument
on four of the five floors of the structure. However, before confronting the
memorial itself, visitors must park or get dropped off at the Visitors' Centre
before walking up the fairly steep path to the base of the monument. Within
the centre there are a host of books about Wallace and Scottish history,
souvenirs and related signs of Scottishness which mediate the site as symbolically Scottish, but also promise further contextualisations about Wallace
following purchase. Most strikingly, adjacent to the Visitors' Centre, at the
edge of the car park, is an imposing stone sculpture of Wallace, a decapi-
tated head at his feet and a large broadsword in his hand. Roaring
'Freedom' – the word inscribed beneath the figure – the sculpture bears the

unmistakable likeness of Mel Gibson in *Braveheart*. This vision of Wallace has been traduced by many locals, who vilify the work as exemplifying cultural mediocrity and banality. Indeed, it has been defaced and subsequently restored. The *Stirling Observer* (Wilson 1997) bemoaned that the memory of Wallace was being exploited, and a local SNP councillor declared that the statue would 'detract from the *true*, very important history which the monument stands for' (my italics).

It is fascinating that in an era of supposed de-differentiation, in which cultural forms merge into one another, and distinctions between high and low culture become blurred, a statue that captures a film-related image of an iconic figure remains highly contested. The outrage seems to be informed by a sense that the earlier tradition of one form of media – the stone monument or memorial, with its conventions of style and representation used to convey the gravity and import of the historical figure – ought to retain its integrity, to stand apart from movie representation. The codes of filmic and sculptural mediation clash here. This is especially surprising given the already numerous examples of more conventional statues of Wallace throughout Scotland, which collectively take no evident common sculptural form, although they do not really divert from conventional memorial appearance. But in addition to the proliferation of other artistic and touristic renditions of the icon, they do add to the melange of stylistic portrayals. Perhaps the movie portrayal is not expected to contaminate other, more serious cultural forms, for fear that Scotland will start to resemble a tartan theme park whose production is disembedded from a local and national context. There clearly remain popular notions about cultural trivia and substance, which appear to differentiate between 'art' (in this case, exemplified by serious memorial commemorative tradition) and the more ephemeral mass culture of film and television. Nevertheless, the *Braveheart* statue has remained, defended by the marketing manager of the local tourist board on the grounds that it is this image of Mel Gibson that most people now associate with Wallace.

Wallace as definitely not Mel

After walking up the wooded hill, visitors enter the monument. Next to the pay desk there is a videotaped preview of the audiovisual display on the first floor which seems akin to a movie preview which selects certain highlights from the dramatic re-enactment played out above. There is also a café and another souvenir shop.

Ascending the stone staircase to the first floor, visitors confront perhaps the most striking display within the monument. The presentation takes the form of an engaging ten-minute show which retrospectively charts the rise and fall of Wallace, dramatising some of the major incidents and conflicts from his life. Present in one corner of the chamber is a life-sized Wallace in the shape of an actor's face projected onto a costumed dummy. Another static, chain-mailed figure stands some ten feet away and in between these two figures, facing Wallace, is a screen, about four foot square, upon which are depicted various maps, paintings and charts. In addition, on this screen, actors playing Wallace's adversaries and allies engage in a retrospective sparring with the more solid figure of the hero.

The presentation commences with an English lawyer reading out the charges against Wallace and his sentence while Edward I of England looks on, and details are furnished about Wallace's subsequent dismemberment. This is followed by the story of Wallace's rise and a testimony from his brother, Malcolm, about his strength and courage. Combining affect, emotion and information, the presentation then focuses upon the Scottish victory at the Battle of Stirling Bridge, and the death of Wallace's close colleague, Andrew Moray, who was mortally wounded. Wallace's political and strategic deficiencies are alluded to, notably by Robert the Bruce with the two men sharing aims but diverging in tactics. Finally, the story focuses upon Wallace's defeat at the Battle of Falkirk and his ultimate capture, due to his betrayal by Scottish nobles. The circular narrative returns again to his brutal execution but then ends with an incantation by Wallace and several other voices of the iconic Declaration of Arbroath, written by Bruce and his court after victory at Bannockburn to assert the right and willingness of Scots to defend their newly won independence. Symbolically, the last few lines are intoned by Wallace alone, to emphasise his crucial role in the wars of independence, although they were composed more than a decade after his death. This is an effective and involving show which skilfully introduces some of the major lineaments of the known history within a dramatic framework. The combination of techniques, the mix of three-dimensional and two-dimensional elements provide three foci for onlookers, who stand and must move eyes and body to follow the dramatic narrative, sharing with the fixtures the cold flagstone floor of the room rather than being enveloped within a plush chair. This physical engagement is not akin to the more static, seated engagement with film and television, but draws the onlooker into a more intimate sequence of action. There is a closer, more interactive relationship between witnesses, simulacra, artefacts and screen than that which pertains in a screen–viewer relationship.

Besides this closer interaction, the audio-visual posits a very different series of aesthetics and characterisation from the Hollywood conventions of *Braveheart*. There is no spectacular and aestheticised action, and more negotiation of meaning and purpose between the characters, although the presentation is heavily slanted towards the unyielding heroism of Wallace. While audience sympathies are clearly intended to be with Wallace, his flaws and poor strategies are pointed out to produce a more ambiguous account than the spotless heroism portrayed in *Braveheart*. It is particularly gratifying to note that the actor playing Wallace has none of the glamorous appearance of Mel Gibson, the Hollywood star of *Braveheart*. With crooked teeth, pasty complexion, jowly face and thinning hair, the actor's appearance gestures towards a kind of medieval verisimilitude devoid of shiny locks and designer stubble. The authenticity of effect means that the glamour surrounding the film star is dispensed with, and by virtue of the animatronic techniques, we gain more sense of the physical presence of Wallace. This mediated experience produces a range of intellectual, emotional and physical responses from visitors not akin to the consumption of a film in a multiplex theatre. Producing different effects, it cannot be subsumed under any general category of sign culture.

On the other side of the room is another, more conventional audio-visual, largely based on drawings and paintings, accompanied by charts and maps, and with a commentary supplemented by sound effects, which tells the story of the Battle of Stirling Bridge, the site of Wallace's greatest victory against his English foes – and again concludes with Bruce's victory at Bannockburn. While this display is informative about the context and tactics of the battle, when the two displays are operating concurrently, there is a rather surreal interaction replete with non-sequiturs, as the two narratives become entangled.

The Hall of Heroes

On the second floor lies the Hall of Heroes, an attempt at the time of the monument's establishment to lend the patriotic allure of Wallace to other famous Scots, each sponsored and championed by individuals and groups. The result is the collection of twelve busts sponsored by these advocates. Again, this is a specific form of nineteenth-century mediation in which the representational techniques of the time – namely the sculpting of the head and shoulders as those deemed worthy of such a memorialisation – were displayed in order to convey a series of ideas about success and status that chimed with contemporary values. The sixteen busts are of Robert Bruce, Robert Burns, Buchanan (poet and Protestant campaigner), John Knox, Allan Ramsay (artist), Robert Tannahill (poet), Adam Smith, James Watt, Walter Scott, William Murdoch (engineer and inventor), David Brewster (inventor of the kaleidoscope), Thomas Carlyle, Hugh Miller (geologist), Dr Chalmers (founder of the Free Church of Scotland), David Livingstone and W.E. Gladstone. Most of these figures – all men – were contemporaries of the monument and supported and served imperial ends through their endeavours in the esteemed realms of literature, painting, scientific invention, Protestantism, statesmanship, economics and exploration. Although there are selections which gesture towards a more exclusive sense of Scottishness, their presence lends allure to the Victorian middle-class values of thrift, industry, religious devotion, enterprise and invention, British qualities which were also imagined to be embodied in Wallace himself. At the ceremony to lay the monument's foundation stone in 1861, Wallace was eulogised as a proto-*British* figure, and upon inaugurating the campaign to erect it, the Earl of Elgin declared:

if the Scottish people have been able to form an intimate union [with the English] without sacrificing one jot of their neutral independence and liberty – these great results are due to the glorious struggle which was commenced on the plain of Stirling and consummated on that of Bannockburn.

(quoted in Morton 1993: 215)

The understanding of Wallace as British now appears laughably outmoded and so do the conventions of representation inherent in the display of marble busts as a means to authoritatively transmit a respect for characters and the values they supposedly embody. Such technologies have been super-seded by other means of celebrating the significance of individuals, whether through the biopic, biography or waxwork.

That the exhibition of busts should be understood as symptomatic of its time is made implicit in the interactive audiovisual presentation positioned in front of it, which draws attention to the historical context for the selection of these men as exemplary Scottish heroes, highlighting the Victorian values that informed their choices, their valorisation of individualism, their mascu-line ideals and their zeal for monuments. This display identifies categories which might be mobilised to make contemporary selections of heroic or important Scots, seven fields identified as 'Liberty and Justice', Sport', 'Art', 'Science and Technology', 'Music', 'Literature' and 'Drama'. A click on any of these designated categories reveals four or five major names accompanied by sound effects, such as applause in the case of drama or the sound of a typewriter for literature. A further click furnishes information about what are often rather unfamiliar characters, and from this cast of thirty or so, visitors are encouraged to vote for the Scot of their choice. While this also might seem to constitute an equally arbitrary and limited selection based on current tastes, each category also contains a more extensive list of other contemporary figures. Thus the data bank teems with over one hundred Scots from numerous periods and fields of endeavour. Contemporary lumi-naries include Jackie Stewart, Jock Stein, Eduardo Paolozzi, Jimmy Shand, Lulu, Muriel Spark, Sean Connery and Billy Connolly, all figures from popular culture, a move which tends to de-auraticise the idea of a hall of fame, again mingling the popular and the 'high'. Accompanying the interac-tive programme is another audiovisual display which details the history of the erection of the monument during the era of romantic nationalism, using old photographs, maps and drawings.

In its previous incarnation, the Hall of Heroes brooks no argument in its choice of marble worthies but masquerades as an 'official' or 'authoritative' selection. Clearly, the present audio-visual complements the Victorian exhi-bition and decentres this power, embodying an interactive form of transmission which avoids the 'expert' and often arid historical tracts included in more traditional guidebooks. Besides providing a contextualisa-tion which undermines the notion that there can be any common-sense, self-evident cast of national heroes, it reveals that such selections are always partial and mediated by cultural values, for the alternative choices are not put forward to evince any substitute preferences. Rather, the sheer number of names and the invitation to consider those who may have been omitted offers a model for subjective choices, cajoling visitors into musing upon the criteria mobilised for such selections. This highlights how official, authoritative versions of history as fact are increasingly undermined both by the proliferating, open-ended versions which populate popular culture and media and by the demise of the expert as communicator. The assembled, self-selected Victorian gentlemen who purveyed their values through the selection of their heroic cast were *legislators* whereas the devisers of the audio-visual interactive display are more concerned with offering *interpretation* and encouraging visitors to similarly interpret historical and contemporary forms of memorialisation (Bauman 1987). The active engagement with the display, the facility bestowed on visitors to operate the system via a menu, rather than through unmediated instruction, and the physical and mental engagements required inculcates an embodied participation in the politics of national identity. Contemporary national identity is thus now revealed to be a mutable entity, constituted out of the jumbles of associations which constellate around ideas of Scottishness but can be assembled in infinite ways, avoiding the fixity of exclusive and defensive identities mobilised around more limited cultural resources (Edensor 2002).

Linking space and time

The monument's third floor features a large-scale panorama which mirrors the view that can be witnessed from the floor above, the viewing platform with its overarching hollow crown. This painted simulacrum of the vista covers all four walls and includes surrounding villages, mountains and other natural features and sites of historic interest, providing a template for what can be witnessed above. The only other feature is an audiovisual display which contextualises some of these mappings by means of diagrams, maps, drawings, photographs and in particular, aerial film. Adopting a historical narrative, it discusses the geological origins of Stirling and surrounds, prehistoric settlement and key archaeological discoveries, Roman colonisation and the later battles between Picts, Scots, Celts and Angles, the medieval military campaigns of Wallace and Bruce, the Scottish Renaissance, the agricultural revolutions of the eighteenth and nineteenth centuries, the evolution of railways and roads, and the rise of Victorian Stirling, and culminates in an overview of the present-day town and its environs. The display thus contextualises the surrounding geography and its history in accordance with a linear narrative and selective sites which move outwards from the monument as a source of orientation, charting a skein of associations that can be first witnessed from the viewing platform and later visited *in situ*. Thus this audiovisual display is a tool that informs viewing but may also lead to an outward momentum towards other possible sights/sites and beyond, whether this is imaginative or actual travel.

This display highlights how the consumption of tourist sites and their attendant visual attractions must always be located within a multitude of trajectories and itineraries which tourists follow, through which they are constantly becoming tourists. Some will travel further, to the far north or the Cairngorms; others are embarking on a day trip, or staying in the nearby Trossachs for the weekend. The experience of sites is thus always shaped and informed by their situation within a wider geographical matrix, of an itinerary or of an individual history of visiting a range of historical attractions. So it is that tourist places, and the stories and images that are disseminated at them, can only be comprehended as enmeshed within multiple flows, events and other spaces rather than as detached sites replete with depthless signs. Like all places, tourist attractions are an ensemble of interlinked, successive experiences and preoccupations that depend on a host of contextualising influences, people carrying in their passions (Scottish history, golfing or birdwatching) traces of other times and places, projects and preoccupations, incompatible understandings that decode sites and their mediated offerings in often contesting ways.

Conclusion

I have described above some of the multiple technologies through which the figure of William Wallace has been mediated, with a specific focus on the new audiovisual technologies installed at the Wallace Monument. I have suggested that this intensification of signs and images surrounding the monument might appear to exemplify the empty signifiers which typify postmodern image culture, disembedding the local specificities and forms of knowledge in the production of yet another super-commodified attraction. However, I argue that a depthless consumption amidst this array of signs is but one way in which such mediations might be interpreted, experienced and utilised.

Too often tourist activity and experience has been synonymous with the tourist gaze, so that tourists are depicted as entirely ocularcentric beings, embodied only through their eyes. This oversimplification neglects other sensual dimensions of being a tourist and dematerialises places so that the concentration on the visual and semiotic features of attractions overwhelms their tactile, sonic and odorous qualities. Places are always characterised by abundant affordances which impact upon the body and cause it to feel and respond in ways other than the visual. If we consider certain modes of tourism, for instance the beach holiday or trips to Ibiza to become immersed in club cultures, it becomes apparent that visual sensations are not even the most prominent senses engaged. The auditory, tactile and olfactory stimulation that overtakes visual experience in such settings indicates that the visual is not always the pre-eminent sense of tourism (Edensor 2003). This is indeed the case at the Wallace Monument. Although this article is not based on an ethnographic project, I questioned fifty visitors on a single November day in 2002 following their visit to the monument, about what they considered to be the 'most memorable' feature of their visit. Despite the panoply of audio-visual installations that may seem to have reconstituted the character of the attraction, 80 per cent of respondents related that the viewing platform at the top was the highlight, 10 per cent cited the spiral staircase that connects the floors of the memorial, 6 per cent referred to the encased broadsword supposedly belonging to Wallace, and a mere 4 per cent identified the drama featuring the animatronic model of Wallace as the high point of their visit.

As emphasised before, the Wallace Monument is itself a form which mediates the significance of William Wallace through the semiotics of its architecture and its location within a symbolic landscape, but meaning is also mediated in the affordances of the structure; in the ways in which it forces visitors to experience a range of different sensations through their engagement with the building. From the Visitors' Centre, a not inconsiderable trek up the rather steep Abbey Craig leads to the attraction (there is a minibus for the infirm and elderly). This is only the start of a physical engagement with the monument for the ascent to the top via the only passage – a narrow spiral staircase – is likely to induce breathlessness and perhaps vertigo. There is no lift. The coercion of the body via the staircase demands a rather unfamiliar bodily movement. Visitors have to squeeze past each other whilst ascending or descending and the tactility of the roughly hewn stone on the ascent are dissimilar to the smooth surfaces of much tourist space. At the apex of the structure, the bracing air, driving rain or warm winds that assail the visitor's body from the viewing platform likewise suggest an engaged rather than a passive body – braced against the cold or scrutinizing the horizons. The view from here is framed by the whalebone arches of the hollow crown, providing an unorthodox border for the spectacular vistas of Stirling and its environs.

Here then, the visitor is not enmeshed in the visual provisions but as Crouch puts it 'bends, turns, lifts and moves in often awkward ways that do not participate in the framing of space, but in a complexity of multi-sensual surfaces that the embodied subject reaches or finds in proximity and makes sense imaginatively' (1999: 12). These spaces might also be described as what Bachelard (1969) identifies as 'felicitous', mysterious, tactile and enclosing, and they act to decentre the visual whilst simultaneously incorporating it.

Having foregrounded the sensual qualities of these highlights, I do not mean to suggest that the visual is detached from the other senses, for it is essential to reinstate the visual as a highly kinaesthetic sense which conjures up non-visual sensations. I have tried to show that the interactive display in the Hall of Heroes, the audiovisual presentation which relates to panoramic Stirling, and the dramatised presentation of the animatronic Wallace engage the body in a range of different ways both *in situ* and also in ways that suggest imaginative and virtual sensations which reach out to other spaces and times. Mediated images depend on the accompaniments of sound and rely for their efficacy upon a (subjective) knowledge of the feel of the world, so that for instance, photographs and paintings of rural landscapes provoke an empathetic response in the viewer, where the affordances of grass, mist and rocks are imaginatively apprehended, perhaps on the basis of fore-knowledge of actual engagement with space. The textures of people, places, atmospheres and occasions are imaginatively reclaimed in fantasies that involve a sensual imagination. This is as much the case with film as with the visual technologies described above. Here, as Barbara Kennedy insists, for cinema-goers, 'the look is never purely visual, but also tactile, sensory, mate-rial and embodied' (2000: 3). The host of visual stimuli at the Wallace Monument may bring forth associations with other media (*Braveheart*, for instance) but also depend upon the recall of other objects, places and times which is informed by tactile, auditory and olfactory memories, by embodied experiences and practical, sensuous knowledge.

The different effects and complex engagements of the audiovisual displays at the Wallace Monument supplement the range of other mediated forms of and about William Wallace. I have emphasised elsewhere that this production of Wallace has always been rooted in popular geographies, rituals, creativities and interpretations but alongside these vernacular constructions, 'expert' and 'authoritative' accounts have tended to predomi-nate at tourist attractions and museums which have been conceived as places of instruction (Edensor 2002). Recent media forms, however, veer away from this legislative approach and offer resources for interpretation rather than instruction. The installations I have examined are more sensual, open-ended mediated forms of transmission than the didactic guidebook or informational display.

This extension of the range of visual mediated technologies and the different effects they may engender has the potential to incite a reflexive engagement with the myth of optical objectivity, with what Martin Jay (1992) has called a modern 'scopic regime'. Jenks argues that this reflexivity contrasts with an 'objective' approach, for 'whereas the dominant visual mode "looks" for the "essential" and the "typical", an interpretive vision pulls, extracts or abstracts its phenomenon into a new setting' (Jenks 1995: 13). This foregrounds a (constantly changing) interpretation of the visual rather than the consumption of an ossified image. This is surely the case with regard to the myth of William Wallace; that he is constantly drawn into new framing techniques which mobilise an array of narratives and images. The figure of Wallace then, unlike the official representations of military heroes, party chairmen and revolutionary leaders, continues to escape attempts to fix his image and meaning. There has always been a proliferation of Wallaces. New mediated forms, including *Braveheart*, mix into a dense matrix or resource bank which reinforces his iconic significance but permits a wide range of appropriation and experience.

References

Appadurai, A. (1990) 'Disjuncture and difference in the global cultural economy', in M. Featherstone (ed.) *Global Culture.* London: Sage.

Bachelard, G. (1969) *The Poetics of Space*, Boston: Beacon Press.

Barnes, T. and Duncan, J. (eds) (1992) 'Introduction', *Writing Worlds*, London: Routledge.

Bauman, Z. (1987) *Legislators and Interpreters*, Cambridge: Polity.

Cohen, A. (1985) *The Symbolic Construction of Community*, London: Tavistock.

Crouch, D. (1999) 'Introduction: encounters in leisure/ tourism', in D. Crouch (ed.) *Leisure/ Tourism Geographies.* London: Routledge.

Cubitt, S. (2001) *Simulation and Social Theory*, London: Sage

Edensor, T. (1998): 'Reading Braveheart: representing and contesting Scottish identity', *Scottish Affairs*, 21, autumn: 135–158.

—— (2001) 'Performing tourism, staging tourism: (re)producing tourist space and practice', *Tourist Studies*, 1: 59–82.

—— (2002) *National Identity, Popular Culture and Everyday Life*, Oxford: Berg.

Edensor, T. (2003) 'Sensing tourism', in C. Minca and T. Oakes (eds) *Tourism and the Paradox of Modernity,* Minneapolis: University of Minnesota Press.

Higson, A . (1995) *Waving the Flag*, Oxford: Clarendon.

Jay, M. (1992) 'Scopic regimes of modernity', in S. Lash and J. Friedman (eds) *Modernity and Identity*, Oxford: Blackwell.

Jenks, C. (1995) 'The centrality of the eye in western culture: an introduction', in C. Jenks (ed.) *Visual Culture*, London: Routledge.

Kennedy, B. (2000) *Deleuze and Cinema: The Aesthetics of Sensation*, Edinburgh: Edinburgh University Press.

McArthur, C. (ed.) (1982) *Scotch Reels: Scotland in Cinema and Television*, London: British Film Institute.

Morley, D. and Robins, K. (1995) *Spaces of Identity*, London: Routledge.

Morton, G. (1993) 'Unionist Nationalism: The Historical Construction of Scottish National Identity, Edinburgh 1830–1860', Ph.D. thesis, University of Edinburgh.

Samuel, R. and Thompson, P. (eds) (1990) *The Myths We Live By*, London: Routledge.

Urry, J. (1995) *Consuming Places*, London: Routledge.

—— (2002) *The Tourist Gaze* (2nd edn) London: Routledge.

Warner, M. (1993) *Monuments and Maidens*, London: Verso.

Wilson, F. (1997) 'A bunch of Wallies?', *Stirling Observer*, 10 September.

9 Mobile viewers

Media producers and the televisual tourist

Robert Fish

Television is tourism

It is much easier to imagine tourism as a category of experience exceeding the formal sphere of television than it is to think of television without the idea of tourism. Travel, visit, movement, escape, displacement, abandonment are all terms that underscore the touristic imagination, but so too are they intimately tied to how television is now understood as a series of encounters between text and a 'watching' audience. Indeed, it is a limited definition of mass media that reduces its touring cultures to one of material flows of people moving from 'screen to scene', although this is how such a convergence has often been broached (See for instance Tooke and Baker 1996; Riley *et al.* 1998). According to writers such as Urry (2002) television is more than simply a non-touristic process anticipating, distinguishing and sustaining material places as sites for a tourist gaze. It is implicated in a series of different tourisms, as much defined by the apparently immobile viewer fixed in front of a depthless screen as it is by locations haunted by the ghosts of a text:

> With TV ... all sorts of places can be gazed upon, compared, contextualised and gazed upon again. ... The typical tourist experience is anyway to see named scenes through a frame, such as the hotel window, the car windscreen or the window of the coach. But this can now be experienced in one's living room, at the flick of a switch.
>
> (Urry 2002: 90–91)

Such a formulation of television is a marker of what Urry, among others, has termed 'post-tourism', a phenomenon defined by a more playful, self-conscious and fluid category of tourist within contemporary life. That is to say, the television text is regarded as a resource for imaginative movement, with the apparently static viewer, by implication, mobile. Such an idea seems reasonable if we were to follow an axiomatic premise of television criticism that the television text must be read in order to be meaningful, since as De Certeau (1984) argues, all readers are travellers actively poaching texts for their own tactical and playful ends. To the extent that we might think of a convergence between television and the touristic imagination in these terms we could therefore do worse than say television viewing is tourism, or better still, a set of tourisms.

In this chapter, I wish to follow the implications of this idea of the mobile audience – the televisual tourist – to shed light on constructions of popular textual content and form. In particular, taking as my focus narrative television texts that foster ideas of countryside as a central feature of their dramatic repertoire, I wish to explore how one specific moment of television, the notion of 'viewing', is actively constituted by media producers through notions of tourism and the tourist, and in so doing, offer insight into how particular notions of countryside come to be circulated. In tackling this issue, the emphasis of this analysis is therefore on reversing the conventional direction of much work into mass media and the tourist imagination. Rather than explore how such media foster or sustain a tourist gaze, as the logic above suggests, it asks: how does the category of tourist sustain the preparation and content of media texts?

Narrative media have an important place within popular cultural imaginings of countryside. Examples include such series dramas as BBC Television's *All Creatures Great and Small* (1978–89), *Hamish Macbeth* (1995–97), *Ballykissangel* (1996–2001) and *Dangerfield* (1995–99), Yorkshire Television's *Darling Buds of May* (1991–93), and *Heartbeat* (1992– present), Carlton Television's *Peak Practice* (1993–2002), as well as long-running continuous dramas such as Yorkshire's *Emmerdale* (1972–present), and serial adaptations such as BBC's *Pride and Prejudice* (1995). In one sense such texts provide exemplary cases through which televisual tourism can be understood in terms of the varied promotions, spectacles and attractions created around screened locations. They are closely implicated, for instance, in the creation of 'new spatial divisions with new place names' Urry (2002: 130) constructs such as *Emmerdale Farm Country, Heartbeat Country, Peak Practice* Country and *Heriot Country* through which the tourist gaze is 'produced, marketed, circulated and consumed' (ibid. 130). Important though these processes might be to new configurations of leisure and movement in the countryside, my purpose here is to move beyond (or indeed back from) the windows and windscreens of Urry's cars and coaches, and consider instead some of the cultural politics of the post-tourist as it relates notions of rurality within everyday life. In particular, taking as my focus three British rural television dramas – Dangerfield, Heartbeat and Peak Practice – I wish to show how media producers, by constructing notions of the televisual tourist, rationalise and reproduce a particular set of textual constructions of rural life.

Running alongside this analysis is a more specific argument about received wisdoms concerning the status of the rural within mass media. In particular, I wish to offer something of an antidote to the notion that mass media constructs of rurality can simply be subsumed under the easily digestible code of a 'rural idyll', arguing instead that such media must also be understood in terms of the production of conflict, since conflict is a defining quality of any narrative form. That is to say, rural texts employing a narrative logic through the mass media must be understood as producing highly ambivalent constructions of rural life; ones that constantly switch between the idyllic and anti-idyllic. This provides a key point of departure for inspecting the idea of televisual tourism, for my argument is that it is in this ambivalence that televisual tourists come to be imagined in different ways and textual themes legitimated. In the final section of the paper I consider what this logic means for the cultural politics of television as it relates to the touristic imagination. I wish to begin, however, by briefly explaining how studies of the rural have accounted for the media.

Rural media equals rural idyll

Trajectories of rural study have for some time now sought to explore the contested and reflexive nature of the sign 'rural' through the terrain of social and cultural theory of which studies of mass media have been an indicative part. As Lawrence (1997: 5) has put it, critical study of the coun-tryside 'understands from the outset that an assortment of communicative forces are brought to bear on the creation varied representations of rural', a comment that has been worked through into empirical insight across a range of media genres (e.g. Brandth 1995; Bell 1997; Phillips *et al.* 2001; Fish forthcoming). And yet, if we were to examine the way in which mass media images of rurality have been typically related to the dynamics of contempo-rary social change in the countryside we would find a fairly predictable reading of rural images as they are seen to service material processes. The particular issue that I wish to elaborate upon briefly in this respect has been neatly summed up by Matless (1994: 8) when he suggests that 'too often images of the country are lumped together under a simple category – 'the rural idyll' – as if there were no difference there, as if the culture of the rural in England were one of a simple sentimental homogeneity'. Notions of idyllic rurality have, of course, had a pervasive role in the way academic research seeks to engage with the signifying power of the rural (e.g. Newby 1980; Mingay 1989; Short 1991; Bunce 1994). While writers such as Cloke and Milbourne (1992: 359) cautioned some time ago about its speculative nature and the need to explore 'the degree to which it is important to repre-sentations of the rural', rural research has readily drawn on this idea to capture the meaning of popular cultural constructions and duly wheeled this idea out as an explanatory variable in understanding the dynamics of contemporary rural change.

Indeed, if mass media texts are imaginative resources that rehearse, define and shape understandings and uses of the countryside, they appear to do so in a fairly uniform and predictable way. So for instance, Bunce (1994:38) has written on the emergence of a countryside ideal in Anglo-American culture, and suggests, in one great sweep, that 'with the arrival of new publishing technology and especially of electronic media – radio, film, and television – in this century, the countryside ideal has been absorbed readily into mass culture'. He writes, for instance, of the way the BBC Radio drama *The Archers* depicts 'the last remnants of a happier way of life' where 'all classes co-exist in tolerant harmony' (ibid. 55). He reflects too on the nostalgia surrounding the long-running BBC TV drama *All Creatures Great and Small*, whose appeal apparently rested on its portrayal of the 'quaint remnants of pre-industrial ways of living' based around 'an old social order' and the construction of a 'benign rural class system' (ibid. 50). Similar conclusions have been drawn elsewhere. Urry (2002: 87) has written of 'the proliferation of new magazines which help construct ever more redolent signs of a fast-disappearing countryside', while Jones (1995: 39) suggests that 'popular discourse plays an important part in the creation and dissemination of the idea of the rural idyll' taking as his exemplar the Yorkshire Television adaptation of H.E. Bates' *The Darling Buds of May*. While exploring the ostensibly different terrain of rural horror and the anti-idyll, Bell (1997: 95) sets up his argument by coding up large chunks of popular cultural form 'from Thomas Hardy and Edward Carpenter to Heartbeat and the Archers' as idyllic constructions of rurality which offer escapist texts for urban viewers. Similarly, Boyle and Halfacree (1998: 308) have drawn attention to the way that 'mass media and marketing agencies' might be critical in spreading a 'generalised belief' that 'pits the "rural" against the "urban"', servicing 'idyllic' representations of rurality which in turn initiate broader processes of social change.

Touristic processes feature explicitly within some of these intersections of media text and social processes. For instance, Butler (1998: 219) writes of the 'great effectiveness' of 'evocative images of rural areas' on processes of rural recreation and tourism, citing such dramatic texts as *Heartbeat* and *Pride and Prejudice*, while Mordue (1999: 631) has written of the way the culture industry working through particular sites 'consistently perpetuates images of an English Rural Idyll', again drawing attention to texts such as *Heartbeat* and *All Creatures Great and Small*. Writing in a Canadian context Hopkins (1998: 65) has also drawn attention to the way that the 'touristic countryside' is replete with 'country ideals which aggrandise myths of rurality' through, among other things, the texts of film, television, books and advertisements. I now wish to inspect such claims through a brief analysis of narrative form as it relates to these rural texts.

A journey through narrative countrysides: idyll and anti-idyll

At some unspecified point in the mid to late 1960s a young man known as Errol decides to takes a trip to the North Yorkshire countryside. It does not start well. His arrival in a small village is met with some degree of suspicion by the locals, not least because Errol is black. With the upbeat sounds of the sixties hit 'Keep on running' ringing in his ears, and against a fetching backdrop of open landscape, Errol's initial experience is one of being chased off some agricultural land by a suspicious farmer and his over enthusiastic dog. We cut to an establishing shot of the village public house and then to a warm, cloying, interior. Errol has arrived to book a room, but despite being empty of guests, its proprietor George refuses to offer Errol a bed. While George apparently has 'nothing against him', it is the attitudes of locals who 'may not be ready for it', and by the look on some of their faces as they drink their beers in silence, he is not wrong in his judgement. Though Errol strikes lucky with a local bed and breakfast who 'welcome the business', no sooner has he made himself comfortable than the local police have implicated him in a burglary at the farm where he was apparently trespassing some days earlier. We then witness a series of serene views over the North Yorkshire landscape as relief from narrative developments. Matters then take a turn for the better when the locals' discover Errol's aptitude for cricket. A grudge match against a rival village is beckoning. Everyone stops looking at the colour of Errol's skin and starts observing his batting technique. Absolved from the crime and with the lively sounds of The Beatles 'Here comes the sun' playing out over the sunlit cricket pitch, Errol single-handedly wins the cricket match for the village. By the close of his trip Errol is so popular that he even contemplates settling in the village for a while.

The village in question, that of *Heartbeat*'s major dramatic setting of Aidensfield, is no stranger to visitors like Errol in the Episode *Giving the Game Away*. Heartbeat's premise has been to explore the life and dilemmas of a village policeman working in North Yorkshire during the 1960s. To do this, its dramatic life has been built around a steady drift of people passing through and arousing suspicion of criminality by dint of their outsider status. Sometimes these narratives end in a positive point of closure, such as the example above, often they do not. There is an account yet to be written of the important narrative purposes the figure of the tourist, traveller and wanderer serves within media constructions of countryside, although my point here is to draw attention to the role of conflict in how they develop meaning for their audiences. Indeed, since narrative is defined by the design of events, and events imply occurrence, change and an altering of states, only in very particular circumstances can the narrative medium function as the didactic system of signification implied by accounts of the rural mass media above. Rather, the process of creating event horizons is typically synonymous with the production of conflict. Events produce meaning by constantly reversing premises, by shifting between counterpoints and creating dilemma. As Robert McKee, seminal practitioner and proponent of story art in film states, 'nothing moves forward in story except through conflict' (McKee 1998: 210). Indeed, this need for dramatic conflict has serviced storylines of incest, child abuse, racism, suicide, and assaults with shotgun and knives in *Heartbeat*. This point is true of the other two series. In *Peak Practice*, whose drama centres on the experiences of group of doctors negotiating their lives in the Derbyshire countryside, rural life is explained through a panoply of illnesses and accidents. In *Dangerfield*, a series based around the life and dilemmas of a police surgeon in Warwickshire, issues of health and crime have become implicated in each other through, among other things, murder, robbery, missing persons and pagan rituals.

All of these themes – event, change, conflict – are developed out of the necessities of form and should alert any analysis from the outset that all is not perhaps well in the rural worlds narrative mass media produce. But neither are these texts exclusively negative as my summary of *Giving the Game Away* begins to show. While the narrative logic of these texts is about expressing dilemma, the meanings are mixed. Narrative events may produce drama and conflict, but the images and sounds constructing them may produce ambivalent readings of rurality: the wide shot of a sunlit Warwickshire landscape where a murder has taken place in *Dangerfield*; the spectacle of the Peak District landscape as a car accident occurs in *Peak Practice*; the cloying imagery of the public house as racial prejudice is enacted in *Heartbeat*. Meanings also change as the text unfolds. Errol's experience of country life is, for instance, qualitatively different at the end of the narrative than at the beginning. Similarly, the act of resolving dilemma is often the means by which more affirmative ideas of rural social practice in the countryside can come in to view. Where conflict is rife, moments of cohesion, consensus and equilibrium will always occur, like a narrative break, and in so doing, a fleeting glance of other, more positive accounts of rural life are developed.

The recurring story-world of each of these dramas offers too a possibly different reading. In all three dramas, individual narratives rely on a construction of community that allows information to circulate and dilemmas to be collectively confronted and overcome. The main protagonists of each of these dramas are all constructed around celebrating the cultural competencies of the professional middle classes in rural life. All are self-actualising, intuitive, motivated, organised, emotionally confident and self-effacing professionals who negotiate rural life with remarkable ease and whose lifestyles and ownership of commodities within them underscore narratives of conflict and dilemmas with a quite different vision and experience of rural life (see Phillips *et al.* 2001). And yet, securing these positive constructions of rural life relies on a litany of demonised others whose experiences of these rural worlds are qualitatively different. Errol, the outsider, in *Giving the Game Away*, is just one instance in these examples of rural drama that consistently demand idyllic lifestyles to be contrasted with figures of suspicion, derision, deviancy and pity. Indeed, to the extent that we can speak of these dramatic forms as framing a reading of the countryside for audiences, it is mixed. Consider, for instance, the following comments:

> [V]iewers complained that an edition [of *Peak Practice*] shown on November 29 was inappropriate for pre-watershed broadcast. The Broadcasting Standards Commission upheld complaints about a scene in which a nurse suffocated an injured marine, which it considered too explicit for pre-watershed.
>
> (*Media Guardian* 20 May 2002)

> Storm in the Peaks! It's chaos in Cardale as Eva Pope revels in the part of TV's meanest Bitch.
>
> (*TV Times*, November 2001)

> 'It's exactly what we major on: the rural idyll. Peak Practice is *Brigadoon*. The leprechaun could pop down the high street at any minute now. Anything could happen in Cardale.'
>
> (Peak Practice Production Team)

> 'Can Will save the village practice? *Peak Practice* coming up next.'
>
> (Voice-over, Carlton Television)

> 'I think on one level [Peak Practice] is quite idyllic. But in a way that warmth, that Oxo family warmth, means that I and the series can tell quite difficult stories. As long as you have that glow to it really, that idyll to it, then you can tell whatever story you want ... the ambulance drives through the countryside which is shot through a golden glow. It's really a strange hybrid.'
>
> (Peak Practice Production Team)

All of the comments above constitute, in one way or another, accounts of the series *Peak Practice*, a drama charting the lives and dilemmas of a group of doctors living in the Peak District, England. One of the comments tracks the complaints of viewers clearly offended by the inappropriate content of its show. Apparently, they were vindicated by the sobriety of a 'standards' committee. No such reading is at work in the magazine the *TV Times*, which playfully headlines village chaos on its front cover. A member of the Peak Practice team I interviewed is equally upbeat, but for different reasons. *Peak Practice*, I am told, is *Brigadoon*. A screen voice-over is perhaps less sure about such a claim. According to this, we should prepare for a village practice in crisis. Another member of the production team is confused. He sees a curious hybridity of idyll and anti-idyll; an odd transaction between narrative events and imagery. Such is the ambivalent architecture of the rural text, spilling out into quite different sorts of reading. I can do no better than the cover text surrounding a recent book production devoted to fans of rural drama: 'If you think nothing ever happens in the countryside, or that nobody's interested when it does, then our TV programme makers are out to prove you wrong. Crime, intrigue, a bit of sex and plenty of comedy – it's all there in copious supply, filmed against peaceful rural backdrops where everybody knows everyone else and anything is liable' Gray (1998).

Travelling audiences/imagining audiences

> Now I know that while I was writing the first script, unbeknown to me they'd got a control group of people all over the country and they're asking them the question: 'A' do you like the title Derfield? 'B' Do you like it being set in the countryside? ... When I started out in television we never considered the audience. We did what we liked, because we thought, if we liked it, everyone else would.
>
> (Derfield Production Team)

It is in the ambivalence that I have just outlined that televisual tourism and ideas of rurality can begin to be understood. In the extended section that follows I consider this premise as it relates to the living-room viewer, a construct no doubt anathema to accounts of tourism predicated on the mobile body negotiating physical space, but one that is integral to the way the idea of tourism asserts influence within the media sphere. I wish to explain this point as it relates to the thoughts and practices of media producers involved in the creation of these three dramas, arguing that notions of the televisual tourist in the production of rural drama are the means by which idyllic and anti-idyllic constructions of rural life are reproduced.

I should begin by stating that constructions of the audience as 'tourist' must be understood as part of the television industry's attempt to overcome its resolutely dispersed, invisible and privatised modes of reception and a means to reproduce its own particular conditions of existence: the generation of income by way of advertising revenue or licence fees. Such a disjuncture between the social relations of production and consumption is what Raymond Williams (1974:18) once termed, television's 'deep contradiction'. And just like any other system of mass communication that operates in multiple time-space formations, it creates a 'distinctive kind of indeterminacy' in which producers lack 'direct and continuous forms of feedback' (Thompson 1995:13) from their audiences. The process of imagining an audience – its needs, identities and desires – therefore becomes central to how the television text is rationalised and prepared. Faced with an increasingly competitive marketplace for audiences, public and commercial broadcasting has increasingly sought to overcome this problem by developing a range of formal techniques aimed quite directly at making the audience less capricious – such as the use of 'setmeters' (registering what output is watched and 'peoplemeters' (registering who is watching it) as well as a more informal set of imaginings of the audience on the part of media practitioners as they reflect on their work (Ang 1991; Brierley 1995; Hagen 1999). The production of these three rural dramas – where I have spent extended time observing the work of media practitioners[1] – is no exception to this process. In particular, what I wish to argue here is how their creation relies on two significant constructions of the audience as tourist: the first by relating notions of an idyllic rurality to that of audience escape, distraction and switching off; the second by relating notions of an anti-idyll to the process by which escapist viewers are educated into the realities of rural life and conflict. Consider each of these in turn.

Idyllic texts and televisual tourists: discourses of escape

> 'You're in the world of Warwickshire. For 50 minutes of a day you've just suddenly *been placed* in that Warwickshire village.'
>
> (Dangerfield Production Team)

According to the first formulation, viewers of rural drama are constituted as wishing to temporarily displace their own conditions of existence; conditions otherwise more real and immediate. They are constructed as escapist viewers. There are precedents for this idea within academic discourse on mass media. In its formative work, modern television criticism often advocated such an account of television. In defining a space for the active audience against the excesses of media determination, the uses and gratifications perspective, for instance, saw in the popular media all manner of ways by which the text might be appropriated for viewer ends, with notions of escapism, distraction and diversion as important among them (see for instance McQuail *et al* 1972). While debates in cultural studies have largely moved on from charting such functional uses of the media, not least for their weak sociological imagination of viewer needs (Ang 1996), the principle of escapism is actively at work in the productions of *Heartbeat, Peak Practice* and *Dangerfield*. According to one member of the Peak Practice Production Team, as long as the drama, 'had that sort of exclusive world to it that is different, that *takes you out* of your everyday life, then it will work'. Similarly, among *Dangerfield* practitioners the audience were given something they could 'switch off with', while in *Heartbeat* output was deemed by one as simply 'escapism'. In short, images of rurality were often constituted by what Richard Dyer (1977) and Christine Geraghty (1991) have termed 'utopian solutions', providing the audience with a set of alternative scenarios of social life to contrast with, and compensate for, the limits of their everyday existence.

That these constructions were thought to allow the viewer to 'switch off' and be 'taken out' of their everyday life means that investment in these escapist worlds was always being tied back to a negative set of spatial and social realities that audiences were trying to transcend. As a set of spatial realities, for instance, a propensity for the countryside was often bound up with the aspirations of the urban dweller and the way that urban life was seen to deprive people of a basic need to experience rural landscapes. As a script-editor on the *Heartbeat* production team suggested:

> 'Well most of the people do live in cities ... whether through choice, or because they were born there. [The countryside] is something they can aspire to ... it's just a little bit more remote; it *takes people away.* That's why these rural dramas are so successful, escapism for the townies.'

In a recent discussion of city in film, Lapsley (1997: 197) has made a similar point, highlighting how both cinema and television have often tried to create 'a space set over against the city'. Media producers of the three dramas regularly constructed this sort of experience onto audiences as a way of rationalising appeal. The spatiality of the text was seen as a distant, purer, forfeited zone of existence arousing and satisfying a very basic set of human needs that their own situated (urban) lives could not deliver:

> [*Peak Practice* audiences] are not in landscape, but it reminds them of landscape ... open space, and mountains and streams.... I think there is a deep psychological need for landscape which is about you and us as complex human beings, and I think this is tapped into with *Heartbeat* and *Dangerfield*, and certainly *Peak Practice.*
>
> (Peak Practice Production Team)

The point I wish to make is the way such constructions of audience function as means for advocating the textual themes produced. The effective needs of the urban audience are not simply a category of reflection for practitioners; one created in hindsight. It is the part of the logic of reproduction by which images of unbroken, open landscapes come to the fore. So for instance, the script editor from *Heartbeat* above was also to add that, 'we *use* the fact that we're in these areas, that we can have people sit back and just lie back and take it in'. Similarly, a script editor on *Peak Practice* who spoke of, 'a massive majority of people who watch the television live in inner cities, and don't see a green field from one day to the next' suggested that, 'so you know, you're looking to create those opportunities'. And this logic works at all levels of production, as a member of the *Dangerfield* production team reflected,

> 'That's where the executive decision-making comes in. It's like [mimicking executive] "no you can't have [this shot inside], it has to have

this rural aspect ... one has to be aware that this the reason it's watched therefore please try and incorporate it".'

Such imaginings of the urban audience were often undercut by compensatory class discourses too, especially in terms of the productions of *Dangerfield* and *Peak Practice*, which apparently offered urban working-class audiences idealised middle-class rural worlds beyond their immediate grasp. Indeed, faced with the inevitability of their own conditions of existence, such imagined audience constituencies could at least engage with their desires through the pleasures of the image. As a producer of *Peak Practice* put it, 'the urban working class have got no chance of having a weekend cottage. They've got no chance of going out there because unemployment is huge, but they use it as escapism'. Or as a sound-editor on *Dangerfield* suggested when clarifying his notion of the urban audience, 'it's for working class people. Yer know, if they are sitting in their two up two down ... they have aspirations'. And again, what we witness in these constructions of the televisual tourist is a logic for re-producing textual themes, ones built around accentuating middle class lifestyles and commodities in their constructions of rural life. In *Dangerfield*, for instance, a location manager, speaking of the way he imagined the audience in these working-class terms, spoke of 'that middle class comfortable bourgeois way of life' that characterised the main protagonist's rural universe and one that, when searching out possible locations, 'is always in the back of your mind'. In the same series, a practitioner involved in production design linked such escapist discourse to a 'glamour element' that they 'certainly don't deny', adding that:

> 'It goes with the way the house is designed and the way they dress. They dress sort of country smart, you know what I mean, "country smart". So it's like, corduroys and moleskin trousers. It's sort of nice. Nice and neat.'

Such connections between the textual form and the televisual tourist continue in other ways: the relationship between urban audiences and constructions of cohesive rural communities in *Heartbeat* and *Peak Practice* as a counterpoint to the alienation of urban life; the relationship between romantic doctors in the countryside and notions of a desiring female audience trapped at home in the imaginations of *Dangerfield*'s producers; and so forth. My point, however, should be clear. By creating the category of the escapist audience, the televisual tourist, media producers are able to make sense of and reproduce a particular version of rural life developed in these dramas. I now wish to reverse this premise and explore an altogether different construction of the televisual tourist and by dint of this, an altogether different mode of constructing rural life in television drama.

Anti-idyll and televisual tourists: discourses of realism

> 'I don't just want people to say about *Dangerfield*: "Oh how pretty, let's all go and see Warwick Castle." I want it to be hard hitting. It has to have that edge.'
>
> (*Dangerfield* Production Team)

If narrative constructions of rurality are understood in one sense by media practitioners as sustaining an escapist touristic gaze, one that effectively allows audiences to depart from the living-room and temporarily suspend their own socio-spatial conditions, then so do they have a second inflection, based around notions of realism. The matter at stake here is for media practitioners to eschew the affirmative aspects of the rural text and to constitute audience interest around its dilemmas: 'imagine seeing just shot after shot of the Peak District', as one member of the Peak Practice Production Team put it, 'you'd be bored silly in couple of minutes'. Indeed, central to the process of rationalising the production of these texts in audience terms is to construct it in the conflictual terms I outlined above:

> 'The basic ingredients are a strong main story which will have genuine problems and dilemmas at the heart of it because otherwise it becomes a bag of sweets. People I think will lose interest.'
>
> (*Heartbeat* Production Team)

> 'You've got to say we can only afford that lyricalness, we can only afford that nice slow pace with all those pretty pictures, if we are going to be brave, and I think when you start to get soft stories and keep all that, it all falls apart and you sort of end up watching TV thinking, what am I watching?'
>
> (*Peak Practice* Production Team)

According to the above comment, lyrical images of setting combined with soft stories were a recipe for disaster, the drama would 'fall apart' in the eyes of the audience. 'Brave' stories, in contrast, acted as a counterweight to these images. They kept people watching. They maintained audience interest. At least part of this process relies on a process of self-identification among media practitioners, with producers of these dramas reluctant to consign their work to discourses of television that appear to denigrate what they consider a creative and powerful practice. Notions of escapism, distraction and 'switching off' are indeed signifiers of these negative commentaries on television, and are duly played down as much as they recognised. Rather, producers wished to think of their dramas as 'thought-provoking' (*Heartbeat* Production Team), and 'hard-hitting' (*Peak Practice* Production Team). The process of escape outlined above, which rests on a compensatory discourse of text–audience relations, therefore becomes reconstructed as a different sort of pact. Watching rural drama may be about inviting certain constituencies of audience to suspend their own negative socio-spatial conditions and enter a highly positive and aspirational set of discourses of rurality, but so too were they about understanding the expectation of conflict. As one script-writer on *Heartbeat* explains it:

> 'What I'm looking to do is take people on a journey … if you look at a production like this there's no denying a certain nostalgia, but what I'm interested in as a writer is taking people on a *journey* … exposing the viewer to some of the realities of life in an out of the way place. There's nothing unusual in that. There's an expectation there [on the part of the audience] which is about being true … it's all a bit sugary otherwise. I think your audience wants you to *take them there*.'

Interestingly, while notions of an escapist audience were about very specific socio-spatial conditions of the audience, the construction here was never about supplanting particularities. It was about realising the effective need of 'the journey'. Journeys were about self-conscious learning and discovery of hitherto unknown lives, and an understanding that all lives are lived in some way through conflict. As a picture-editor on *Peak Practice* eloquently put it, 'television drama is all about expressing the dilemma of life. Otherwise it's false. Audiences understand that. They relate to that'. And yet, some media practitioners recognise that these journeys run in contradistinction to, or at least in an antagonistic alliance with, other tourisms informing the text. Indeed, if the idea of conflict is constituted as a necessity of the form, one apparently understood by producers and viewers alike, it can also imply a largely condescending view of audience when constructed around discourses of escape. When the dramatic worth of the text is constructed around its ability to challenge, rather than pacify, the audience, the escapist viewer becomes simply an object for instruction. The idyllic text is something to be supplanted with a less than positive account of rural life for the unsuspecting viewer:

> 'You've got a programme which is set in the kind of nice and beautiful surroundings, about a handsome doctor who lives in the country and you take that formula and seduce your viewer *into that world* and then you can explore really meaty issues which have some kind of social conscience'.
>
> (*Dangerfield* Production Team)

Much like those whose antidote to superficial, disengaged acts of mass tourism is to assert an equally problematic notion of real travelling, so we find media practitioners often wanting to recast their work as magnifying the realities of rural life for an ignorant and unwitting escapist audience. As the member of the *Dangerfield* Production Team above put it, through the 'nice and beautiful surroundings' the (escapist) viewer becomes 'seduced' in to alternative set of rural realities. Like the characters in these dramas, the televisual tourist is taken on voyage of self-understanding and transformation, 'allow[ing] people to absorb the rights and wrongs of a life in a community like Aidensfield', as one script editor on *Heartbeat* put it. Or as another on *Peak Practice* explains, 'people think the countryside is just this total ideal. Yer know, no illness, no pain, no sadness or whatever. In our own small way we're saying, life's just not like that, *wherever you go*'.

Conclusion

In this discussion I have sought to use the category of the televisual tourist as a means to explain how notions of rurality in popular television drama are informed. In developing this account I have also questioned a reading of the rural media within academic discourse whereby ideas of the countryside are simply subsumed under notions of an idyllic rurality, and against which wider processes of rural change, such as tourism can be explained. The account I provide here argues that such readings are at best partial. Narrative media allow highly ambivalent constructions to emerge, which rest, in a significant sense, on how the television text is constructed as an aesthetic resource for the imaginative movement of an immobile viewer. Such notions of (im)mobility are conditioned by their own cultural politics. Televisual tourists are variously seen as figures to be compensated by rural images because of their negative social–spatial positionings or otherwise ignorant publics needing to be educated into a particular set of rural social realities.

These categorisations on the part of media producers demand inspection, but they do beg questions of the simple transaction that apparently operates between televisual countrysides on the one hand and a tourist imagination on the other. Indeed, despite the sense in which the post-tourist is meant to capture a more self-reflexive subject, one that might use the television text both tactically and playfully as tourism, it is telling that one of its advocates, John Urry, also suggests in a related discussion that, 'only certain sorts of countryside are attractive to the prospective visitor' (Urry 2002: 88). As my discussion has made clear, it is less than straightforward what constructions of rural life dramas such as *Dangerfield*, *Peak Practice* and *Heartbeat* appear to foster, or what kind of visits they encourage. It is precisely in this antagonism, between the semiotic potential of the text and the tactics of the reader, that we can begin to explain how the media, as part of the touristic imagination, comes to take shape, and assert influence.

Notes

1. Between 1995 and 2000 I interviewed and observed more than fifty media practitioners involved in all stages of the production process in the three dramas.

References

Alasuutari, P. (ed.) *Rethinking the Media Audience*, London: Sage.

Ang, I. (1991) *Desperately Seeking the Audience*, London: Routledge.

——(1996) *Living Room Wars: Rethinking Media Audiences for a Postmodern World*, London: Routledge.

Bell, D. (1997) 'Anti-idyll: rural horror', in P. Cloke and J. Little (eds) *Contested Countryside Cultures: Otherness, Marginalisation and Rurality*, London: Routledge.

Boyle, P. and Halfacree, K. (1998) 'Migration into rural areas: a collective behaviour framework?' in P. Boyle and K. Halfacree (eds) *Migration into Rural Areas: Theories and Issues*, Chichester: John Wiley.

Brandth, B. (1995) 'Rural Masculinity in Transition: gender images in tractor advertisements', *Journal of Rural Studies*, 11: 123–133.

Brierley, S. (1995) *The Advertising Handbook*, London: Routledge.

Bunce, M. (1994) The Countryside Ideal: Anglo-American Images of Landscape, London: Routledge.

Butler, R. (1998) 'Rural Recreation and Tourism', in B. Ilbery (ed.) *The Geography of Rural Change*, Harlow: Longman.

Cloke, P.J. & Milbourne, P. (1992), 'Deprivation and Lifestyles in Rural Wales II: Rurality and the Cultural Dimension', *Journal of Rural Studies* 8, pp. 359–372.

De Certeau, Michel (1984) *The Practice of Everyday Life*, trans. S. Rendall. Berkeley: University of California Press.

Dyer, R. (1977) 'Entertainment and Utopia', *Movie* 24.

Fiske, J. (1987) *Television Culture*, London: Routledge.

Geraghty, C. (1991) *Women and Soap Opera*, Cambridge: Polity Press.

Gray, H (1998) *TV Country Favourites*, Skipton: Atlantic.

Hagen, I. (1999) 'Slaves of the Ratings Tyranny? Media Images of the Audience', in P. Alasuutari (ed.) *Rethinking the Media Audience*, London: Sage.

Hopkins, J. (1998) 'Signs of the post-rural: marketing myths of a symbolic countryside?', *Canadian Geographer* 80: 65–82.

Johnson, R. (1986) 'The story so far: and further transformations', in D. Punter (ed.) *Introduction to contemporary cultural studies*, London: Longmans.

Jones, O. (1995) 'Lay discourses of the rural: developments and implications for rural studies', *Journal of Rural Studies* 11: 32–49.

Lapsley, R. (1997) 'Mainly in cities and at night: some notes on cities and film', in D. B Clarke (ed.) *The Cinematic City*, London: Routledge.

Lawrence, M. (1997) 'Heartlands or Neglected Geographies? Liminality, power and the hyperreal rural', *Journal of Rural Studies* 13: 1–17.

McKee, R. (1998) *Story*, London: Methuen.

Matless, D. (1994) 'Doing the English Village, 1945–90: an Essay in Imaginative Geography', in P. Cloke, M. Doel, D. Matless, M. Phillips, and N. Thrift, *Writing the Rural: Five Cultural Geographies*, London: Paul Chapman Publishing.

McQuail, D., Blumler, J.G. and Brown, J. R. (1972) 'The Television Audience: A Revised Perspective', in D. McQuail (ed) *Sociology of Mass Communications*, London: Penguin.

Mingay, G. (1989) (ed.) *The Rural Idyll*, London: Routledge.

Mordue, T. (1999) 'Heartbeat Country: conflicting values, coinciding visions', *Environment and Planning* A 31: 629–646.

Newby, H. (1980) *Green and Pleasant Land? Social Change in Rural England*, Harmondsworth: Penguin.

Phillips, M., Fish, R. and Agg, J. (2001) 'Putting together ruralities: towards a symbolic analysis of rurality in the British mass media', *Journal of Rural Studies* 17: 1–27.

Riley, R., Baker, D. and Van Doren, C. (1998) 'Movie Induced Tourism', *Annals of Tourism Research* 25: 919–935.

Short, J. (1991) *Imagined Country*, Routledge: London.

Tooke, N. and Baker, M. (1996) 'Seeing is believing: the effect of film on visitor numbers to screened locations', *Tourism Management* 17: 87–94.

Thompson, J. (1995) *The Media and Modernity: A Social Theory of the Media*, Cambridge: Polity.

Urry, J. (2002) *The Tourist Gaze: Leisure and Travel in Contemporary Societies* (2nd edn) London: Sage.

Williams, R. (1974) *Television: Technology and Cultural Form*, London: Fontana.

10 'I was here'

Pixilated evidence

Claudia Bell and John Lyall

Abstract

The tourist has more photo opportunities than ever before. Relentless documentation accompanies every touristic experience. Where once the tourist
with the 35 mm camera sought to emulate *National Geographic*, now with
digital video the tourist engages television's *Lonely Planet*.

Images can even be made and transmitted when weather prevents the backdrop from matching the tourists' expectations. At Jungfraujoch, Switzerland,
when the mountain is obscured by cloud, tourists pose with a choice of pixilated backdrops to photograph themselves, then instantly email those images to
friends around the world. The attached email might say 'here we are at the
mountain (that we did not see). But here is evidence that we were there': an
integral relationship between tourism and mediation, as well as between tourism
and place. We also note the use of pixilated evidence on a passenger ferry at Ha
Long Bay, Vietnam, and for visitors to ancient cave drawings in Zimbabwe.
Because of the inter-compatibility of digital technologies, tourists' own
images, Web-sourced images and fantasy images are all in the same form.
Boundaries are blurred. Any digital image can be printed, downloaded,
emailed or posted on the web, to become part of infinite pictorial digital
space. There is no need of particular technical photographic expertise; therefore there are even more photo opportunities. And photography can never
be exhausted. There will always be something else, or somewhere else, to
photograph; always another image bite. 'We exist in a media saturated environment, which means that life is quintessentially about symbolisation,
about exchanging and receiving – or trying to exchange or resisting reception of – messages' (Poster 1995: 26). 'The camera and tourism are two of
the uniquely modern ways of defining reality' (Horne 1984: 22). Pursuing
these ideas, we present a series of narratives of contemporary tourism.

Cell Phone Man: digital im*media*cy

In Vietnam a Londoner named Nick, 35, told us of his adventures. He likes
to visit war zones – while there is a war on. 'It is the ultimate adrenalin rush,'

he said. He told us he'd been in a situation in the Lebanon with bullets whizzing past. 'I phoned home on my cell phone so they could hear too, and said, guess where I am!' It will be even better, he conceded, when he gets a cell phone that transmits pictures.

Nick's story illustrates the point that where once one took the photograph to be consumed later at leisure, one can now take the photograph with a cell phone and email it to friends as an act of instantaneous contemporaneous and gratuitous oneupmanship. Is it now the case, through that digital immediacy, that the only convincing image is the one that is both made and consumed by its target audience in the present? The currency of the image now decreases almost exponentially with longevity, because we can make and receive these messages in the present. A current television advertisement for picture-taking cell phones has a mountain bike rider speeding downhill through city traffic at 60 kph. He has strapped his cell phone to the handlebars, to take a picture of his face, to send to a friend similarly engaged on another street in another city. At the same time he receives the adrenalin-activated portrait of his friend, also hurtling down hill. These images have not just im*mediac*y; but an im*mediat*e digital and fully congruent and overlapping conversation; to some extent a simultaneous collision of image bites. This is a far better chance to emulate the emotion, the kinaesthetic, the adrenalin-fuelled downhill ride. In this exchange, two simultaneous monologues rather than a conversation, each player has to have the current currency. The life of the exchange is measured in nanoseconds. (Reminder: a picture is worth a thousand words!)

This technology allows Nick and the mountain bikers to say not just 'I was here'; but 'I am here, right now, having this experience in real time, and here is the evidence that this is the case. ... (and, in Nick's case, you are not!)' (And unless you can send your competing narrative, that will remain the case.) Both sender and receiver are using the same technologies; they belong to a club. They are technically literate producers and consumers of 21st-century narratives. The receiver implicitly understands the im*mediac*y of the 'story.' This impresses the receiver; why else would the caller phone them? They are sufficiently part of the same tribe, who recognise the consumption of risk activities (not just sports; now also war!). The audience has enough overlap with some of those experiences to be an informed consumer of the message.

But Nick the tourist has swapped television genres. While the eco-tourist might be re-enacting *National Geographic* or David Attenborough, and the extreme sports junkie is re-enacting A. J. Hackett or Davo Karnicar (Bell and Lyall 2002); now Nick has upped the ante for the adrenalin junkies. He has real time evidence that he is acting out hard news war coverage. On his vacation he has joined that elite club which includes Sean Flynn and all those news photographers who died in Vietnam and other more recent wars, like Iraq. This closing of the gap then continues. Having got himself into the same situation as contemporary war photographers, Nick now needs to stay there to maintain the rush. It is not at all difficult to find out where to go: 'just look at television news!' There are also web sites for this new category of travel, 'danger zone tourism', aimed at travellers planning visits to (or returning from) dangerous places (Adams 2001: 266).

Nick told us he really hopes to become a freelance war photographer. He showed us his brand new professional camera gear, and the instruction booklet. Art imitates life? In the war in Iraq, photographers are now at even greater risk (and more obviously complicit), as they are 'embedded' in military units. They are immersed in a sea of media (media sea; immediacy); but have lost their self-sufficiency. While they have greater technological freedom than ever before, they have traded away their autonomy.

Mediation at Ha Long Bay, Vietnam

Ha Long Bay is a United Nations Educational Scientific and Cultural Organisation (UNESCO) World Heritage Site. Its three thousand small islands (marine invaded tower karst) cover an area of 1500 square kilometres in the Gulf of Tonkin off the coast of North East Vietnam. Comparable geological formations may be seen at Guilin in China, and Krabi in Southern Thailand. This is a totally iconic landscape for the Vietnamese. Numerous ferries take visitors from the mainland out to see the magnificent karst islands, about 3 kilometers offshore. The magnificence of Ha Long Bay attracts people. Its positioning as a UN Heritage Area underpins its profile and status, and therefore expands visitor interest.

Lonely Planet's join-the-dots tourism, its removal of the joy of serendipitous discovery, is aimed at the insecure traveller who will check out *Lonely Planet* television programmes, pore over its web sites, fax email-less hotels, email the rest, leave nothing to chance. A quick hunt for Ha Long Bay on the World Wide Web reveals numerous sites. Tourist and media are heavily intertwined even before one leaves home. *Lonely Planet* was once a guidebook but is now often encountered as pixels on the screen; both the television show and the web page: pixilated. In other words the tourist production of pixilated evidence 'I was here' is conducted according to the pixilated manual, which of course is constructed from the evidence assembled by other voices, other tourists, other travellers.

For the visitor, the scale of the landscape at Ha Long Bay is so large and so immersive that it defies capture, even with a medium format panoramic camera. From the front deck of the boat, we took our pictures, needing multiple frames to reassemble as a joiner photograph later. Everyone else was inside, most of them talking, a few sleeping, because we had not yet arrived at the designated spot from which to take *the* photograph.

A group of young men joined us at the front of the boat; not to feel the salty spray and the soft wind, as it turned out, but to get better reception for their cell phones. They proceeded to send text messages. These messages may not even have addressed the business of gazing at landscape. They were nevertheless sent from a boat floating in the panoramic midst of these World Heritage-listed islands. The severely diminished vocabulary of texting, sent out to assemble itself on the screen of another digital phone as pixels, had even less chance of capturing the moment than a spoken conversation. The rapturous uptake by the early adopters of the new digital technologies sometimes leaves them less able to respond than far more traditional technologies.

The craze for cell phones now invades any tourist space. It is now also part of extreme sports spaces. Mountaineers and yachties are armed with cell phones to send out an urgent distress call, should this be necessary; and to effect an urgent rescue should this be necessary; and to stay in touch with home, to counter the loneliness. The dangerously isolated adventurer is becoming a rarity. As New Zealand mountaineer Rob Hall lay dying near the summit of Mount Everest in 1996, he was comforted by the sound of his wife's voice, half a world away. On September 11th workers in the World Trade Center, or on the plane that later crashed near Philadelphia, phoned home to speak to their loved ones in their last moments. Those poignant messages that some of them left that day on answer-phones are evidence of how even the most traumatic, the most tragic and the most personal events are now mediated.

The young men at Ha Long Bay with their cell phones are part of contemporary populist consumer culture, responding to commercial demands to participate in the latest technological novelties. But of course, cell phones are not a complete novelty. The cell phone has evolved by drawing from, reconfiguring and recombining elements of earlier media. It was not invented out of nowhere.

Goode explains that 'it is tempting to think of the mobile as 'telephony plus', the 'plus' signifying a growing series of augmentations such as mobility (framed not only by the emergence of wireless networks but also by progressive miniaturization), SMS 'txting', MMS 'pxting', WAP, email, voicemail, i-mode and so on.' He tells us that this hierarchical model cannot explain the (unanticipated) rise of text messaging as 'killer app', which has driven sales of mobiles in many countries. Many people use mobiles almost exclusively for texting (Goode 2004). Nor does the technology itself explain why users are so anxious to upgrade to newer models, when the advances are not in increased performance or function, but in aesthetics: colour, size, ring tone. Maffesoli's tribes are at loose here: the tribal consumer needs to be armed with the appropriate object, in this case the cell phone. As Maffesoli explains, there is not such a concern now with individuals and their social positions, but 'roles acquired through the reversible practices of self-fashioning, roles which enable the feeling of participating in a general representation' (Maffesoli 1996). Digital communication of shared experience provides tribal membership. The tribe assembles in cyberia; but is a community, nevertheless, of persons who in the real world might be isolated individuals.

Cellphone technologists keeps searching for new innovations, Goode tells us: 'moving images, hi-fidelity sound, seamless integration with other gadgets, wearable, invisible and voice-activated devices, and other spectacular developments.' The rage for txting was unexpected. Perhaps, he suggests, 'this "revolution" has already run its course, and successful future developments will be incremental (better interface design, cheaper international communication etc.)'; evolutionary rather than revolutionary. Inevitably some unpredictable new craze will emerge as everything becomes smarter, smaller and faster.

The young men at Ha Long Bay with their cell phones just wanted to communicate with someone, probably to report what they were seeing, where they were. That day they were adding another experience to their lifetime collection. They wanted to tell their friends.

Jungfrau, Switzerland: trans-cyberian im*media*cy

At the Jungfrau in 2001, we found more loci for communication. Jungfrau is an extremely popular tourist site in Switzerland. Even tourists with just three days in Switzerland ('doing' Europe in three weeks') knew that Jungfrau was a 'don't miss' attraction.

In 1898 the Jungfrau Railway up to the Jungfraujoch in Switzerland was opened. This was a significant moment in the history of alpine tourism. The railway carriages take passengers up as far as the Eiger Glacier, to the Eiger Wall Station 3,550 metres above sea level. The route upwards culminates in a tunnel through the Eiger and Monch mountains. Each year over half a million people take this trip. The railway makes this spectacular part of the mountains accessible to non-climbers.

At the top on a clear day one can marvel at the view. But at any of the designated look-out points it is very likely that one might not see much at all, especially in winter. White-out conditions are frequent. Weather can be checked on the mountain surveillance screens on the web site at the railway station at the bottom, or on one's own computer. But mountain weather can change unpredictably. Still, one can explore an ice cave with ice carvings, buy souvenirs, drink schnapps or hot chocolate, wander through the museum display, collect information leaflets; or purchase a postcard, write a note on it and post it (possibly to one's home address: the Jungfraujoch has its own postmark, noting altitude as well as date: a unique memento). Or one can phone home. Visitors are advised that is 'the highest point on earth' from which to make land line phone call. There is also an email booth: one can send an email, and include a photograph taken in the photo booth, which offers a choice of Jungfrau landscapes to use as backdrop, including seasonal variation (alpine flowers for spring; clear blue sky and white glacier for winter, and so on). Such is digital technology: you can be photographed with a view you never saw, and have the image transmitted in real time, while you are still (not) there. If one cannot be immersed in the snow on the mountain itself, one can be far more comfortably immersed in its far less threatening representations.

Even if one cannot see it, on this day trip the Jungfrau is nevertheless a constant presence. Visitors know what they have come to see, and they see it over and again, on their train tickets, on railway station murals and posters, on postcards, calendars, key-rings and other souvenirs, on their restaurant menus and place mats. There is plenty to write home about, even without a snow-capped mountain visible. Many tourists posed next to the backlit mural of the mountain that they never saw. These will be displayed with their other holiday snaps: (the depiction of) happy tourists against (the depiction of) a glorious mountain background (Bell and Lyall 2002; 24–26). From a production of culture perspective of media analysis, we can see that even though this tunnel was built over a century ago, today it is a site that very emphatically produces cultural products and services with new media. The mountain experience is not just a nostalgic recapitulation of late nineteenth-century tunnelling genius; those aspects are juxtaposed against ubiquitous use of the newest communications technologies.

In a sophisticated country such as Switzerland such connections are hardly surprising. What about sites in remote Third World countries? How do tourists there participate in global connectivity?

Matobo, Zimbabwe

We use as our example one of the sites of the oldest evidence of human habitation; remote, authentic, and in Zimbabwe, a country currently considered difficult for travellers.

At Matobo in southern Zimbabwe, visitors are taken by four-wheel-drive vehicles 45 km to the south of Bulawayo, to see the famous Motobo Hill cave drawings. A not-too challenging walk up a short track leads to the caves. Here tourists can marvel at ancient drawings of giraffes, buffalo, impala and hunters painted by the San hunter-gatherers around 20,000 BC. Tourists look at these images, then turn to face their travelling companions, asking to be photographed in front of these ancient drawings. Once again the little cameras click as visitors take turns to stand in front of these delicate masterpieces. In this juxtaposition of technologies, various things are happening simultaneously as the digital cameras consume these cave drawings.

First, the tourist gains a representation of the ancient artefacts. These are a rare sight; indeed, the tourist is shown brown patches on the walls where some of this collection of remarkable drawings was obliterated in the 1920s when local 'conservationists' decided paraffin oil would save the works. Those images rapidly disappeared. But we can still see giraffes, antelope and human figures in various iron oxide ochres and red clay prancing daintily along the cave walls. Every visitor to Zimbabwe sees these animals in the present; so can readily recognise these depictions of the same animals from the past.

The picture of themselves and companions is an instant depiction of temporal bandwidth: the present ('I am here now! in this digital image of me, here in the present and in the presence of these ancient images'). The generic human past becomes backdrop to the personal present; thus under-pinning a fairly anonymous moment with the cultural credibility of an acknowledged global treasure.

Second: here is yet another group of exotic foreign invaders consuming yet another precious, sacred resource of indigenous people. It is a wonder-fully ironic rerunning of Plato's Cave, where the people gather excitedly to peer at the captured digital image, which is important because it contains them, meanwhile largely ignoring the physical reality of the images in front of them. In the digital image the string of binary information, which defines any one pixel, is of exactly the same bit length; a unit of image from the depiction of the tourists is of the same mathematical binary gravitas as a unit of image of the cave drawings.

Third: while they have photographs of the local wildlife, and of the wild life depicted in the past, then of themselves in front of the depiction – what they have neatly done is substituted themselves for the Zimbabwians. They are the humans in the present grounded in their neo-tribal human generic global past (a popular culture understanding of Africa as the Cradle of Humanity).

Fourth: whatever tourists might see, they always require a picture of themselves as well: not as a mnemonic of what they looked like on that day (though years later they might giggle about 'that awful jacket, that ridicu-lous haircut!') but as evidence that yes indeed, 'I was here'. This place may not even be remembered as Matobo, but 'somewhere in Zimbabwe'. In producing that great theatrical drama that is our own lives, we are the central figures – even when juxtaposed against these rare ancient art works in the caves. Plus, we suggest, current popular archaeology gives us so-called known ancestors: Lucy, Eve. In popular imaginings 'we' all came out of Africa. Are some of the people in these drawings our own ancestors? The tourists seeking documentation of their own autobiography in front of the cave drawings appear oblivious to any ironic reading as they pose.

Bibliography

Adams, Kathleen M. (2001) 'Danger-zone tourism', in Peggy Teo, T. C. Chang and K. C. Ho (eds) *Interconnected Worlds Tourism in South East Asia*, Oxford: Pergamon

Bell, Claudia and Lyall, John (2002) *The Accelerated Sublime; Landscape, Tourism and Identity*, Westport, CT: Praeger.

Goode, Luke (2004) (forthcoming) 'Keeping in (and out of) touch: telecommunica-tions and mobile technocultures', in C. Bell and S. Matthewman (eds) *Cultural Studies in Aotearoa New Zealand*. Melbourne: Oxford University Press.

Horne, Donald (1984) *The Great Museum: the Re-presentation of History*, London; Pluto.

Maffesoli, Michel (1996) *The Time of the Tribes*, London: Sage.

McNair, Briar (2002) 'New technologies and the media', in Adam Briggs and Paul Cobley (eds) *The New Media: An Introduction*, Harlow, Essex: Pearson Education.

Osborne, Peter D (2000) *Travelling Light: Photography, Travel and Visual Culture*, New York: Manchester University Press.

Poster, Mark (2002) 'Culture and New Media', in Leah Lievrouw and Sonia Livingstone (eds) *The Handbook of New Media*, London; Sage.

11 'I'm only here for the beer'

Post-tourism and the recycling of French heritage films

Phil Powrie

When is Provence, favoured holiday destination for both the French and British, not Provence, but a heavily mediated cultural appropriation by the British of something which is no longer a topographical space, but a fantasy? The answer lies in the twin manifestations of French heritage films and British advertisements for lager in which, paradoxically, not a word of English is spoken.

Jean de Florette and *Manon des Sources* (France, Claude Berri, 1986) launched the contemporary heritage genre in French cinema. Not only were they successful films in France, but they were popular during the late 1980s in the UK, when they were amongst the few French films to be found in video stores during the couple of years when distribution problems caused the withdrawal of most foreign-language titles. They remained popular in the UK long after the late 1980s; during the 1990s they were frequently screened on the digital channel Film Four, unlike many of the other French heritage films. During the late 1990s the images and the music at the core of the 1986 films were recycled in a series of television advertisements for the (originally Belgian) lager, Stella Artois. These advertisements, set in rural Provence, and whose dialogue was completely in French, were made for the British market (they were never broadcast in France). Just as the feature films launched the heritage genre in France, so too the 'Reassuringly Expensive' campaign for Stella Artois, which began in 1981, established the Lowe Lintas advertising agency as a major multinational company.[1] The advertisements based on the French films, however, did not start until almost a decade after, in 1995. There have been seven such advertisements: 'Carnations', 'Monet', 'Red Shoes', 'Good Samaritan', 'Jacques', 'Last Orders', 'Returning Hero', and 'Doctor', the last two moving away from *Jean de Florette* to evoke, respectively, *La Vie et rien d'autre* (France, Bertrand Tavernier, 1989) and *Le Hussard sur le toit* (France, Jean-Paul Rappeneau, 1995). The two advertisements I shall concentrate on are 'Red Shoes' and 'Last Orders' (The titles are those given by the agency).

These advertisements will help us to understand some functions of the postmodern tourist gaze, as analysed by John Urry. More specifically, I shall analyse how the advertisements construct the tourist gaze through a complex layer of intertextual allusions. These allusions combine, to recall Urry's analysis of the 'prefiguratively postmodern' aspect of tourism, 'the visual, the aesthetic, and the popular' (Urry 1990). The layers of intertextual allusions function as a screen, with the dual function of a screen; they both hide and show. They hide the real tourist space of the South of France through their representations of it, while those representations construct an idealised tourist space of consumption, itself dual in that the high-art allusions legitimise the beer, and the beer popularises the high art. Put more simply, you do not have to travel any more to be a tourist; you need only drink the beer. This will conjure up the 'ideal representations' of the formerly auratic Romantic view of the landscape (Urry 1990: 86), without the inconveniences of physical displacement, in a process of what I shall call 'dislocation'. In that sense, the advertisements are a good example of what Urry calls post-tourism, and I shall be arguing that they construct a post-tourist gaze. Before doing so, we need to review the heritage film and its functions during the 1980s and 1990s, as well as sketch out the narrative of the two French films which serve, at least apparently, as the referents of the Stella Artois advertisements.

Heritage cinema emerged both in the UK and in France during the 1980s. It is a term used to describe films such as *A Room with a View* (UK, James Ivory 1986), in the early 1980s, and *Sense and Sensibility* (USA/UK, Ang Lee, 1995) in the late 1990s. In France, as in the UK, they cover a variety of films, going from adaptations of Marcel Pagnol's Provençal-based novels to literary classics such as *Germinal* (France, Claude Berri, 1993), based on a Zola novel, or on works by less well-known authors, such as, for example, the regional novelist, Jean Giono, for *Le Hussard sur le toit*. Diverse though they may be, these films have common features. As Ginette Vincendeau explains, heritage films, whether UK or French, 'are shot with high budgets and production values by A-list directors and they use stars, polished lighting and camerawork, many changes of décor and extras, well-researched interior designs, and classical or classical-inspired music' (Vincendeau 2001: xviii). As she goes on to point out, the difference between heritage films and the costume dramas of the 1950s and 1970s which preceded them in France is that 'there is a change of emphasis from narrative to setting' (Vincendeau 2001: xix). These films attempt to be historically and topographically accurate. The two key elements in Vincendeau's analysis, for our purposes, are the importance of setting and classical music, as we shall see below. I would add a third element, which is implied by her characterisation of the heritage film: the reliance on (high) art, either through the inclusion of art works in the décor, and/or intertextual allusions to specific paintings, often associated with the cultural heritage (see Powrie 1997a).

If such films are called 'heritage films', it is indeed mainly because of their links with the rise of the heritage industry during the 1980s (see Higson 1993). For that reason they have often been seen as ideologically regressive, and sentimentally nostalgic, by both UK and French commentators. That said, as Vincendeau again reminds us, there is a difference between the UK and French heritage films; the latter are frequently dark in tone, whether it be the melodramatic but ultimately tragic conflict between father and son in *Jean de Florette* and *Manon des Sources,* or the demise of the miners in *Germinal,* or the raven-pecked plague victims in *Le Hussard sur le toit,* or, finally, the Protestant corpses and multiple assassinations in Chéreau's version of the sixteenth-century St Bartholomew's Day Massacre in *La Reine Margot* (France, Patrice Chéreau, 1995).

The narrative of *Jean de Florette* and *Manon des Sources* recounts the clash between wily peasants (Le Papet, played by Yves Montand, and his nephew Ugolin, played by Daniel Auteuil), and the young professional from the big city, Jean (played by Gérard Depardieu), who comes back to his roots in the country, intent on returning to the land in the ecological sense. For this, however, he needs water in the arid countryside which is Provence. Unbeknown to him, his property does have a spring; but his neighbours covet the spring because they are cultivating carnations for sale to city-dwellers. They hide it from him, as a result of which he exhausts himself cultivating crops and rearing rabbits, failing for lack of water; he eventually dies, leaving his daughter Manon. She, however, has discovered the secret of the spring, and eventually takes her revenge on the neighbours and the community at large by blocking the spring. Unbeknown to all of them, though, Jean is the lost and only son of Le Papet, as is melodramatically revealed at the end of the second film; Jean's father dies of a broken heart, realizing that he has killed his only son. The films, then, pit peasant against city-dweller – an issue clearly of interest to us since it constructs the problem of the tourist invasion – but also father against son.

Jean de Florette and *Manon des Sources* can be seen as a dual nostalgic return (see Powrie 1997b). First, they are a return to the golden age of French cinema, in so far as they are identified with Marcel Pagnol's films of the 1930s, set in Provence. Indeed, Pagnol himself made a film called *Manon des sources* in 1953, starring his wife, subsequently writing a novelised version on which the two modern films directed by Claude Berri are based. There is a second nostalgic return, which is the return to the land, to a rural authenticity, much trumpeted in both Pagnol's novel, and in *Jean de Florette* by the Depardieu character. This return to the land is doubly emphasised by Depardieu's well-documented real-life attachment to the French countryside (he owns a vineyard) (see Vincendeau 2000: 233). The discourse centring on the return to the land leads to a comical misunderstanding when Jean/Depardieu says to his neighbour Ugolin/Auteuil that he wishes to 'cultiver l'authentique' ('cultivate authenticity'). His illiterate neighbour interprets this as cultivating a particular plant to rival his own cultivation of carnations, the imaginary 'lotantique'. The films can be seen then, in the French context, as a phantasied attempt to re-imagine the link to the land and village community which has been lost since World War Two, due to the rural exodus to France's major cities during the 1960s and beyond. In this sense the films provide 'visual sites for national identification' (Esposito 2001: 14).

The films and their translation into TV advertisements clearly mean something different in the British context, however. Here, I would like to identify two strands of an argument. The first strand has to do with the meaning of Provence for the British; the second has to do with the displacement from feature film to TV advertisement.

First, the meaning of Provence. The Mediterranean region generally has always attracted tourists. It is, as Urry reminds us, 'the world's most successful destination', with tourist numbers 'predicted to rise from 100 million in 1985 to 760 million in 2025' (Urry 1994: 174). The key issue, however, is where tourists travel in the Mediterranean, and even more, what type of tourists, and what type of tourism. In the nineteenth century, the French upper classes were drawn to Spain, while the British upper, and increasingly middle classes were drawn to Italy (as we can see in the novel and heritage film, *A Room with a View*) and the French Riviera (see Pemble 1987). The rise of the jet-set in the late 1950s and 1960s in France led to the privileging of the French Riviera as the more attractive location for both the French and the British upper and middle classes, thanks partly to Brigitte Bardot's fondness for Saint-Tropez. Meanwhile, Spain, formerly the principal attraction for the French aristocracy's Grand Tour in the nineteenth century, became the prime attraction for lower middle-class and, increasingly, working-class French and British families, thanks to the rise of the package holiday.

By the 1980s, the French Riviera had developed its tourist industry to attract not just jet-setters but also middle-class families, reaching saturation point. It is at this point that British tourists began to 'discover' the countryside beyond the Riviera, both further out to the West in the hinterlands of Marseilles, and further inland towards the Lubéron hills and the Rhône Valley (Avignon, St Rémy-de-Provence, and so on). In these areas the French had for some time been buying ruined farm-houses (the *mas* or *bastides* as they are known locally) as second homes. The British middle classes began to do the same thing, as can be seen in the extraordinary success of Peter Mayle's autobiographical narratives *A Year in Provence* (1989), which spawned two sequels (in 1991 and 1999), as well as being serialised for British TV (with the actor John Thaw transferred from Oxfordshire's verdant pastures in the Morse series to the arid Provençal hills). The British went westwards and inland, where they joined the French middle classes in buying permanent homes, or second homes, or renting *gites*, arguably the ersatz form of the second home for many British tourists.

Figures for this phenomenon are not available, and much of the evidence is anecdotal. Keith Hoggart and Henry Buller, who have investigated property purchase by Britons in France, point out that some '2000 French homes were sold to Britons in 1987, rising to 4000 in 1988, and to 14,000 in 1989, with 20,000–25,000 expected in 1990' (Hoggart and Buller 1995: 182). A more recent survey has added that 'some newspapers have speculated that as many as 200,000 Britons now own properties in France' (Gallent and Tewdwr-Jones 2000: 75). The boom in property purchase in France occurred largely as a result of 'rapid increases in British house prices and the relative ease of obtaining credit in the late 1980s' (Hoggart and Buller 1995: 186). This was at the time that *Jean de Florette* and *Manon des Sources* were distributed in the UK, becoming perhaps the best-known French films, and fuelling to some extent the fantasy of rural home ownership which is so popular among the middle classes, who are statistically the most likely class to own a property in the South of France (see Hoggart and Buller 1995).

Provence, therefore, became a middle-class British colony.[2] There, the issue became integration (or not) into the community, as the sketches by Mayle of curious local customs and types suggest. Provence became at one and the same time a second home, so familiar but also foreign, a country peopled not by the recognisable sixties jet-setters, but by rustics with funny accents, quaint customs, and a fearsome attachment (a very French cliché) to the land. It is not difficult to see how the films *Jean de Florette* and *Manon des Sources* were ideally placed to typify the new tourist environment for the British, with their wily peasants pitted against the urban newcomer seeking 'authenticity', a key notion for the tourist gaze, as Urry has shown. Moreover, the films' status in the UK would have legitimised the middle-class colonisation represented by second homes and *gites*. For if they were resolutely *popular* films in France, the simple fact that they are subtitled films makes them, in the current state of distribution and reception of foreign-language films in the UK, high-brow 'art' films.

They are even more obviously 'art' films, by virtue of their general emphasis on nature as 'landscape', the Romantic ideal which was transformed during the nineteenth century into 'leisure and pleasure – tourism, spectacular entertainment, visual refreshment' (Green quoted in Urry 1994: 175). They are art films, too, by their more particular configuration of that landscape on nineteenth-century painting, more pointedly, recognisable paintings by Cézanne of the Provençal landscape (Esposito 2001: 22).

As I pointed out above, it is not only the art references which make these films relatively high-brow, but also the music, in this case a version of Verdi's overture *La Forza del destino*, arranged for orchestra and mouth-organ. The combination of classical music (to which I shall return), and Impressionist (or more accurately post-Impressionist) *mise en scène* imparts the high-cultural values corresponding to middle-class tastes and life-styles. As Vincendeau says, talking about heritage generally:

> Arising at the same time as the expansion of museum culture and theme parks in the 1980s, heritage cinema has been linked to retro fashion, interior decoration and tourism, provoking derogatory comments such as the 'Laura Ashley school of filmmaking', 'The Merchant-Ivory

'Furniture Restoration' aesthetic'; [Merchant Ivory's] stately brand of adaptation', the 'white flannel' school.

(Vincendeau 2000: xix)

In this section, we have explored the context for the French films, and seen how they function rather differently for French and British audiences. For the French, they function as a fantasied nostalgic return to both the Golden Age of French cinema and to the notion of a cohesive rural community. For the British, they function as a high-cultural signifier of tourist colonisation. For both, they signify a general notion of 'authenticity', but that authenticity is configured differently: for the French, it is the authenticity of their 'roots', for the British, it is the authenticity of the tourist experience, unadulterated by the 'framing' of the 'package'. For both, then, it involves the same kind of authenticity at another level, that of 'ecological' authenticity, a return to an (apparently) unmediated version of landscape, unmediated because 'lived in' (however inauthentically) rather than 'viewed' through the window of the 'tour'. However, unlike the French, for whom owning a second home is not a class issue, that authenticity is specifically class-based for the British, because it is in fact mediated through a high-cultural framework which is more pronounced than it is for the French. Arguably, the Verdi opera and the Cézanne references would be no less high-cultural references for the French. But the status of subtitled films in the UK ensures that high-cultural references are brought to the fore. This is even more pronounced in the advertisements, one might argue.

Turning, then, to the second strand of my argument, the displacement from feature film to TV advertisement, we need to consider how a series of TV advertisements where not a single word of English is spoken could be so successful in the UK, where resistance to speaking in other tongues seems to be an inalienable right. Is it not just a manifestation of what Urry calls 'aesthetic cosmopolitanism', which he characterises, amongst other things, by '*openness* to other people and cultures and a willingness/ability to appreciate some elements of the language/culture that one is visiting' ? (Urry 1994: 167; his emphasis). The answer lies more in the advertisements' postmodern playfulness, which, amongst other things, includes a jostling host of intertexts, both British and French. This, crucially, is what makes the 'Reassuringly Expensive' campaign different from, say, the 'Papa–Nicole' Renault Clio advertisements, whose non-verbal interjections redolent with innuendo and Gallic shrugs inscribe a purely gestural Frenchness. I shall be arguing that these intertexts position the consumer/spectator of the advertisements in a distanced, playful, and ironic tourist gaze, although I shall also claim that this is counter-balanced by the function of music, which reinscribes affect, and language, which ensures 'authenticity'.

Lurking palimpsestically, almost archeologically, under the advertisements is *Jean de Florette* and *Manon des Sources* of 1986, then Pagnol's *Manon des Sources* of 1953, then 1930s Pagnol more generally, and this cinema context is doubled by the social context of the rural exodus. A French viewer would immediately have recognised these allusions (had the advertisements been screened in France), but arguably a British viewer would be unlikely to recognise them. I would suggest that the process I am alleging does work for many British viewers, however, partly because the red shoes themselves reference not a French film, but the Powell and Pressburger classic of the same name of 1948 (where they are metaphors for desire, which, as we shall see, plays an important role in the narratives of the advertisements). The reference to a 'classic' British film with high-art connotations (ballet) encourages viewers both to accept a high-art context, and, I would suggest, to search for the type of complexities which the notion of 'high art' might entail, but anchored in a French context. Here we can distinguish between two types of viewer. Over the last twenty years one of the staple texts on British A-Level syllabuses for French has been Pagnol's autobiographical novel *La Gloire de mon père*, itself adapted to the screen along with its companion volume *Le Château de ma mère* (France, Yves Robert, 1990). 'Pagnol' therefore has been in circulation as a signifier of a particular type of Frenchness (combining childhood, and therefore nostalgia, and the South of France, a favoured tourist location) for some time for the first type of viewers, those British who studied the (elite) French A-level system, even if what he signifies is hazier than it might be for the French. For a British viewer who recognizes 'Pagnol', then, a complex signifier is evoked which combines what Pagnol's narratives are known for (childhood and nostalgia) with 'Frenchness', integrated with the viewer's own subjectivity, itself predicated on notions of childhood (because the majority of such viewers would have studied the Pagnol texts in their teens). For the second group of viewers, those who had not studied Pagnol, the advertisement's 'Frenchness', coupled with a nostalgic rurality, signifies something much less precise, but no less powerful in that its Provençal location would be likely to appeal to many socially-aspirational viewers eager to emulate the well-off middle classes who had been buying properties in Provence, as detailed above.

The temporal complexity of the intertexts (pre-war with Pagnol's 1930s films; post-war with *The Red Shoes* in the case of one advertisement; 1980s with the references to *Jean de Florette* and *Manon des Sources*) leads to 'constructed undecidability', as one theorist of the post-modern puts it (Calabrese 1992: 139), creating the typically postmodern atemporal moment of consumption. The haziness is important, because it acts on the temporal level, much like a cinematographic filter might do on the visual level, to distance and to obscure the object. The object is *dislocated* from its origin, inhabiting the atemporal moment of fantasy, which, following Urry, I shall be calling the post-tourist gaze.

The decentreed origin is crucial for the creation of the distanced spectator/consumer, unencumbered by location or affiliation. The spectator/consumer can say 'this is not me, nor is this my home, although I recognize both'. The result is the impression of freedom to choose, and distance from the characters in the advertisements, who are no more than types become stereotypes. They are peasants who need to be seen, but also need to be displaced by the British spectator/consumer in the colonisation and consumption of both land and lager. This distancing is all the more pronounced by the narratives, which speak of betrayal. The grandmother believes the son has bought her the shoes she so desperately needs, and the father believes he will get the beer the son has been sent for. In both cases, the younger generation betrays the older generation, or origin, filling the gaps created (the hole in the shoe, the emptiness of the glass) by trickery, by a lure, which almost inevitably raises the laughter of knowing recognition in screenings of the advertisements to an audience.

Moreover, distance is created by irony in this pastiche, since the source of desire in the films is, as the title of Pagnol's novel underlines, the 'water of the hills'; in the advertisements, the source of desire is the fermented water of the town, to which the hero must travel to purchase the lager.

Clearly, however, postmodern distance and irony notwithstanding, the spectator/consumer must be hooked somehow. Music in the advertisements functions to do precisely this. It is a commonplace of film music criticism, as evidenced by the very title of Claudia Gorbman's *Unheard Melodies*, that classical film music is subliminal (Gorbman 1987). Her argument is that it eases anxiety, and allows us to be manipulated. Robynn Stilwell even suggests that sound, more generally, 'forces a surrender of control; we cannot turn away' (Stilwell 2001: 171). It is in this way that the spectator/consumer is hooked in the advertisements, abdicating the control which might have been suggested by knowledge of the intertextual allusions, and also abdicating the distance of irony, which, through music, is replaced by the immediacy of affect. It might be argued, of course, that the music is so portentous, so melodramatic, that it too forms part of the distancing which creates the tourist gaze; but, in my view, the arrangement for orchestra and mouth-organ contains the music's otherwise melodramatic excess, the mouth-organ functioning as a kind of extension of the breath of the hero, returning us to the immediacy of a man's body and the travails we see him go through visually.

The music's position in the advertisements' mini-narratives is crucial. In both cases, the pattern of the advertisement is the same. The older generation manifests a desire for a consumer object (red shoes, a glass of Stella; note the clear gender divisions which associate the grandmother with costume and the father with the more manly beer). The younger generation attempts to fulfil that desire by undergoing a task or set of tasks, involving a journey which is for us no more than a tour around Provence, the whole set to Verdi's portentous music for *La Forza del destino*, suggesting tourism as the destiny of the disembodied gaze. But the most recognisable theme, from the overture, already used as the main theme of the French films from 1986, and used to signal the catchline, as it were, the moment of the lure, occurs at

exactly the same point in the two advertisements: when the younger generation tricks the older generation, validating possession of the consumer
object as well as possession of the environment, which the young men had to
tour through to obtain the object of their desire. The origin is thus wrested
from an older generation and consumed in a pseudo-epiphanic moment
engineered by music as much as by visual images, which celebrates youth
over old age, transitory object over more permanent object (the red shoes),
life over death (in 'Last Orders'), immediate pleasure over generosity, and so
on. In other words, there is a playful role reversal compared with the original films. In the films, Ugolin tries to prove his love for Manon by
undergoing a series of trials, but eventually commits suicide, outwitted by
the daughter of a city man. In the advertisements, the Ugolin equivalent
undergoes a series of trials for the older generation of rural types, outwitting
their representatives so that he can get to the object of his desire; not a
woman but a beer.

Music may be important, but so too is language. The fact that only
French is spoken confirms two things. First, it confirms cultural as well as
financial capital, since an understanding of the French language suggests
middle-class acceptance of education, as well as, quite possibly, the practice
of the language associated with frequent holidays in France. Second, and
perhaps much more importantly, it confirms the authenticity of the tourist
experience, however virtual that experience might be given that we are
talking about TV advertisements. It is to that virtuality that we must return
to conclude.

I began by suggesting that an analysis of the interplay between *Jean de
Florette* and *Manon des Sources* and their recycling in the Stella Artois TV
advertisements could help us understand the post-tourist gaze. In the first
section I argued that *Jean de Florette* and *Manon des Sources* underscore
British middle-class colonisation of Provence during the late 1970s and
1980s, connoting a fantasied return to an 'authentic' tourism informed by
high-cultural and ecological awareness. In the second section I argued that
the TV advertisements of the 1990s, a decade later, show that Provence has
become more distanced in the middle-class imaginary, subject to the post-
tourist gaze, caught in a web of playfully constructed intertexts. Let me be
clear that in both cases we are dealing with distanced representations typical
of the postmodern. As Urry points out when talking about travel
programmes on TV, 'the typical tourist experience is ... to see *named* scenes
through a *frame*, such as the hotel window, the car window or the window of
the coach. But this can now be experienced in one's own living room, at the
flick of a switch; and it can be repeated time and time again' (Urry 1990:
100;his emphasis). What is interesting about the 'replay' represented by the
TV advertisements is that they multiply the 'frame', so that what was once a
film on the big screen becomes the film condensed on the small screen, but
constantly referring outwards to other big-screen experiences, imploding
and exploding at one and the same moment. There is a shuttling to and fro

from big-screen experience to small-screen experience, just as there is a shuttling to and fro between the types of experience provided by the various
frames:

> The post-tourist is freed from the constraints of 'high culture' on the
> one hand, and the untrammelled pursuit of the 'pleasure principle' on
> the other. He or she can move easily from the one to the other and
> indeed can gain pleasure from the contrasts between the two. The world
> is a stage and the post-tourist can delight in the multitude of games that
> can be played.
>
> (Urry 1990: 100)

In other words, spectators of the TV advertisements can delight in the references to iconic films on the one hand, with all the cultural capital which it
entails. They can also delight in the references to bodily pleasures, imagining
that they might be quenching their thirst with a lager which is both a necessity, like water in drought-prone Provence, but also a luxury, because it is so
expensive. The one delight rubs off on the other, of course, the high-cultural
allusions legitimising the consumption of lower-class beer, which might
otherwise have seemed, in comparison to the more obvious wine (copiously
consumed by Jean in *Jean de Florette*), too close for comfort to the image of
the lager lout.[3] And, finally, at the antipodes of the image of the lager lout,
they can delight in the lateral thinking and wit of the peasant attracted to
the town, who has managed to outwit those who have stayed in the country,
thereby legitimising their own fantasied position as both part of that
Provençal community and yet distanced from it. The term I used above was
'dislocated'. Music re-establishes the missing affective link, thereby hooking
the consumer to the product (the landscape, the lager). The radically
distanced spectator/consumer can thereby *relocate*, colonising the place and
consuming the product, while still remaining outside both.

The 'authentic', as was the case with the films, has turned into its muti-
lated, scratched mirror-image, the imaginary *lotantique*.[4]

Notes

1. The agency was originally called Lowe Howard-Spink. The 'Reassuringly Expensive' campaign took both the product and the agency into the superleague: Stella
   is now the fourth largest grocery brand in the UK, and the agency is ranked
   fourth among world-wide agency groups, with billings of more than US$12
   billion and offices in over eighty countries.
2. Although it is important to note that the middle-class invasions of rural areas
   throughout parts of Europe 'ha[ve] more to do with opportunity structures than
   innate attributes of a particular segment of the middle classes' (Hoggart, Buller
   and Black 1995: 216).
3. I am grateful to Margaret Montgomerie for raising the issue of the lager lout.
4. I am grateful to Felix Thompson for his careful reading of an earlier version of
   this chapter.

References

Calabrese, O. (1992) *The NeoBaroque: A-Sign of the Times*. Princeton, New Jersey: Princeton University Press. Originally published in Italy as *L'età neobarocca*, Roma. Laterza, 1987.

Esposito, M. (2001) '*Jean de Florette: patrimoine*, the rural idyll and the 1980s', in L. Mazdon (ed.) *France on Film: Reflections on Popular French Cinema*, London: Wallflower Press, pp. 11–26.

Gallent, N. and Tewdwr-Jones, M. (2000) *Rural Second Homes in Europe: Examining Housing Supply and Planning Control*, Aldershot, Burlington, USA, Singapore and Sydney: Ashgate.

Gorbman, C. (1987) *Unheard Melodies: Narrative Film Music*, Bloomington: Indiana University Press.

Higson, A. (1993) 'Re-presenting the national past: nostalgia and pastiche in the heritage film'. In L. Friedman (ed.) *Fires Were Started: British Cinema and Thatcherism*, Minneapolis: University of Minnesota Press, pp. 109–129.

Hoggart, K. and Buller, H. (1995) 'British Home Owners and Housing Change in Rural France', *Housing Studies*, 10: 179–198.

Hoggart, K., Buller, H. and Black, R. (1995) *Rural Europe: Identity and Change*, London, New York, Sydney and Auckland: Arnold.

Pemble, J. (1987) *The Mediterranean Passion: Victorians and Edwardians in the South*, Oxford and New York: Clarendon Press.

Powrie, P. (1997a) '*Un dimanche à la campagne*: nostalgia, painting, and depressive masochism', in P. Powrie, *French Cinema in the 1980s: Nostalgia and the Crisis of Masculinity*, Oxford: Clarendon Press, 38–49.

——(1997b) *French Cinema in the 1980s: Nostalgia and the Crisis of Masculinity*, Oxford: Clarendon Press, pp. 50–61.

Stilwell, R. (2001) 'Subjectivity, gender and the cinematic soundscape', in K. J. Donnelly (ed.) *Film Music: Critical Approaches*, Edinburgh: Edinburgh University Press, pp. 167–187.

Urry, J. (1990) *The Tourist Gaze: Leisure and Travel in Contemporary Societies*, London: Sage Publications.

—— (1994) *Consuming Places*, London: Routledge.

Vincendeau, G. (2000) *Stars and Stardom in French Cinema*, London and New York: Continuum.

——(ed.) (2001) *Film/Literature/Heritage: A Sight and Sound Reader*, London: BFI.

12 'We are *not* here to make a film about Italy, we *are* here to make a film about ME...'

British television holiday programmes' representations of the tourist destination

David Dunn

> Television is an invention that permits you to be entertained in your living room by people you wouldn't have in your home.
>
> Sir David Frost

Introduction

It is, perhaps, less than scholarly to use in a title words which are unattributed and which may be anecdotal, but in the absence of significant academic comment on television holiday programmes they do offer some insight into the genre and a starting point for this chapter. They were quoted by travel journalist Paul Gogarty writing in *The Daily Telegraph* about the appointment as presenter of ITV's holiday programme *Wish You Were Here?* of Anthea Turner, who in 1997 replaced the long serving Judith Chalmers and her co-presenter, travel correspondent John Carter:

> [U]nlike John [Carter] Anthea will not write or research her own scripts. She will sit smiling on a restaurant terrace in front of a delicious sunset. The director will tell her to take a sip of wine, encourage her to smile wider and say something interesting, such as: 'Hmmm ... that tastes good ... in paradise the food and drink are as memorable as the sunsets.' Anthea's appointment is just another stage in the process known as 'dumbing down' that must reach its nadir soon. Once, the location was the star, now it is just an interchangeable backdrop.
>
> (Gogarty 1997: 13)

He added from his own experience of presenting BBC1's *Holiday* that

> [o]n another travel show a female 'personality', when requested by her director to say something about her location, was heard to explode: 'We are not here to make a film about Italy, we are here to make a film about ME!'
>
> (ibid.)

This chapter is concerned with the ideologies and signifying processes of British terrestrial television holiday programmes, that is programmes which offer reports and information about the tourist product and the tourist destination, as evidenced over the last decade. The chapter suggests that they have drawn conventionally on an essentially scopic discourse of place which set certain store by the eighteenth-century traveller's discourse of 'pleasurable instruction' (Batten 1978) in representing the destination as a *locus* where the high-cultural practices of gazing on buildings and other works of art combined, albeit uneasily, with those of the pleasure-seeking tourist. It further suggests, however, that Turner's appointment was just one example of a perceptible shift towards an increasing privileging of celebrity and show business. After consideration of the conventions of the genre as they existed through much of the 1990s the chapter identifies a number of its discursive axes. That these axes reflect the limited and limiting models of tourism circulating within the academy at the same time suggests, perhaps, that holiday programmes have been reactive rather than proactive in their engagement with the practices of tourism. If the academy has subsequently concerned itself increasingly with the limitations of the scopic in accounting for tourist behaviour, and has signalled a shift towards consideration of other senses (Franklin and Crang 2001; Chaney 2002; Crouch 2002) and of performance (Crang 1999; Rojek 2000; Edensor 2001; Coleman and Crang 2002), so too have broadcasters seemed increasingly constrained by their primary role as providers of surrogate sightseeing. The chapter, therefore, concludes by arguing that response to place is being replaced by an increasing foregrounding of the performance and celebrity of the presenter which, in turn, throws the tourist destination into a background of soft focus.

Conventions of television holiday programmes

Traditionally holiday programmes have been shown in the winter half of the year, when audiences were likely to be thinking about the next summer holiday, and when film reports from sunny destinations offered escapist fantasy even for non-holiday-makers. The two major and long-established offerings of the genre are *Holiday*, transmitted on BBC1 in early weekday evening peak time, and ITV's *Wish You Were Here?*, similarly occupying a peak time slot in mid-week. A third programme *The Travel Show*, of similar format to *Holiday* and *Wish You Were Here?* but featuring more upmarket destinations, ran on BBC2 in spring and summer until 1999. An analysis of viewing figures and audience demographics at the height of the genre's popularity in the mid-1990s (*The*) *Times*, 8 January 100595: 21) revealed that *Holiday* had an audience composed of 48.3 per cent ABC1s compared with the national average of 44.7 per cent for all television programmes. At that time both programmes remained regularly in the middle of the top one hundred most popular on terrestrial television, attracting audiences in 1998 of around 9 million, although *Holiday* was the more consistent performer in the ratings. By 2002 *Holiday* had dropped to an average of 6.5 million while *Wish You Were Here?* no longer featured in the top one hundred.[1]

Holiday's more consistent performance may have owed something to the fact that it maintained, until the late 90s, a clearly stated link with the travel industry, crediting travel writer Desmond Balmer as a consultant, while *Wish You Were Here?* had, by 1998, dispensed with travel writers Perrot Philips and John Carter as, respectively, scriptwriter and co-presenter. By 2002, however, *Holiday*'s links with the industry were less formalized, with travel journalist Simon Calder offering only occasional contributions, while *Wish You Were Here?* currently features short consumer spots from resident travel 'expert' Russell Amerasekara, a former travel rep and holiday industry executive.

Notwithstanding the decline in viewers, which perhaps says as much about terrestrial broadcasters' declining share of the audience in the face of competition from satellite and other broadcasters as it does about shifting viewer preferences, the *Holiday* 'brand' has engendered a number of spin-offs on BBC1, including *Holiday Memories*, *Holiday Celebrity Memories* and *Holiday Swaps*. Both *Holiday* and *Wish You Were Here?* follow a similar format, offering a magazine style programme of filmed reports of tourist destinations each presented by a different member of a loose 'team' of presenters drawn more from show business than the travel business. Both programmes fill a thirty minute slot, and for many years the tradition was to have three reports per show, one featuring a long-haul destination, one filmed in Europe, and one in Britain. Since 1995 *Holiday* has featured four or five shorter reports, the extra ones normally featuring a second European destination and a low- cost holiday, and resulting in a faster show with more pace and variety, but perhaps an increasingly superficial one. *Wish You Were Here?* has maintained the three-report format, often giving extra weight to the long-haul destination which is placed in the middle of the programme and divided in two parts by the centre commercial break. This doubtless takes into account the programme's need to hold its audience's interest during the break and the financial considerations, in an increasingly stringent financial climate of competition in ITV, of having to send a crew to an extra location for each programme.

There are parallels between promotional videos and other forms of advertising and holiday programmes (Morgan and Pritchard 1998). There is, however, a need to distinguish between the production determinants of commercials and holiday programmes and to recognise a paradox. Television draws for its reports on traditions of reportage and documentary 'objectivity'. It is, at the same time, dependent on good relations with the travel industry, since the destinations featured are ones offered by specific tour operators, and the practice is for each report to finish with a caption which names the operator and indicates the cost. It is fair to assume that such named operators are likely to facilitate travel and offer other help to the production team, and that there is at least some sharing of interests. This raises issues of editorial freedom. Seaton (1989), in a series of interviews with leading newspaper travel editors, reported that they did not in the main accept facility trips or 'freebies'. Given the costs of filming abroad it may be disingenuous to claim that similar integrity reigns everywhere in television, and the *Guardian*'s Travel Section has asked:

> should TV travel shows be production-line tour-operator propaganda, or should they try to peel the pre-packaged veneer off holiday resorts and give you useful information and a real flavour of the place? The latter option looks less likely by the hour since it's a nonswerving rule that if a travel programme visits a place you've been to yourself, it bears no resemblance to your own experience.
>
> (Sweeting 1997: 19)

While *Holiday* and *Wish You Were Here?* do offer sporadic consumer information such as a costing comparison between a package deal and a self-arranged holiday or an item about the customer care policies of low-cost airlines, their focus on consumers generally takes another direction. There has been a shift in leisure programmes from instruction by a few experts to 'makeovers' whose focus is on ordinary people and the diversity of popular taste (Ellis 2000). The 1997 series of *The Travel Show* featured a new strand *You Pay; We Say* in which viewers turned over their money and invited the programme makers to change their holiday habits, the results of which were filmed and shown. One couple featured (*The Travel Show* 30 June 1997) was sent flotilla cruising in Greece having told the producers of a preference to escape from other people when on holiday. This format has since been expanded on BBC1's *Holiday Swaps*, where participants exchange holidays and are subsequently observed defining their own cultural practices and preferences generally to the disadvantage of the couple whose holiday they have inherited, while *Holiday* (10 February 2003) has featured a *Holiday On The Case* item in which presenter Craig Doyle consulted with a 'team of travel experts', travel journalist Simon Calder and travel psychologist Robert Bor, to profile viewer Aileen McLaughlin and match her to a destination before sending her on a filmed skiing holiday in Sauze d'Oulx. Yet whatever the apparent democratising suggested by such public access to the screen, the balance of power between broadcasters and audience remains weighted, and the price of brief celebrity is the willing submission to the programme and destination priorities of the broadcasters.

The primacy of the scopic

In the construction of narratives of place there have been parallels between the scopic, especially the photographic, practices of the tourist and those of the television camera. Broadcasters have used power/knowledge strategies to their own advantage in gazing on and constructing an Other, while often, through linguistic or other inadequacy, failing to effect an adequate translation of cultures beyond a recycling of familiar images. They have thus supported MacCannell's (1976) contention that people do not see a place but a succession of sights, authenticated by being reproduced. There has in addition been an increasing convergence between broadcasters and the tourist industry in the representation of the tourist destination as commodity, as a place to be consumed. In this they have reflected Urry's (1990) theory of tourism as a scopic activity whose gaze is ordered to facilitate consumption, and his later contention in *Consuming Places* (Urry 1995) that tourist interaction with place reflects the ambiguity of that title. Places are sites of consumption. They can be consumed. They can consume people's identities. They can be exhausted by use.

The primacy of the scopic is contested by many, not least by MacCannell's (2001) claim that tourists reject the institutionalized gaze for something less visible and more empathic; and while Urry's (2002) own revision of *The Tourist Gaze* maintains that sight is viewed as 'the most discriminating and reliable of the sensuous mediators between humans and their physical environment' (2000: 146), he recognizes a multiplicity of other sense-scapes in tourist behaviour (Franklin 2001: 123). The close relationship between photography and tourism, however, suggests that the scopic continues to figure in holidaymaking. Osborne (2000: 72), drawing on MacCannell's (1976) theory of sight sacralization, writes that '[t]ourists and their sights exist in order to be photographed; indeed are photographed in order to attain their existence'. Crang (1999: 252), reflecting the increased interest in the performative nature of leisure activities, describes tourist photography as 'the capitalizing of experience', adding that constructing pictorial narratives 'is a form of self-creation that is based around a fractured and presentational existence'. In the case of television programmes, which offer viewers surrogate sightseeing, that self-creation has been undertaken by television presenters and reporters. They are not tourists, but they have assumed touristic roles which inform their, and their camera's, narratives and performances.

One cannot, however, take the parallels between the holiday photographer and the television holiday programme camera too far. The holiday photograph is both a more portable and a more democratised artefact than the institutionally filmed or videotaped report, and is a personal production, part of an individual's social system of networks, narratives and displays (Franklin and Crang 2001). The television camera may reflect the iconography of the holiday photograph but it is also the mediator of a performance about tourism staged for viewers, rather than itself being a part of touristic performance. If the use of photographs is 'social evidence of achievement, but pleasurable evidence' (Crang 1999: 249), the representations of the television camera offer evidence rather of the status of the reporter who stands posed in front of it at the centre of its focus, and whose words and persona reflect the ideologies of broadcasting institutions and their creative practices.

Edensor's (2001) use of the metaphor of performance in analysing how tourism is directed and choreographed does, however, offer parallels with the staging of holiday reports both in its discussion of how key workers, directors, stage managers and performers, produce the tourist experience, and in its identification of enclavic tourist space, managed and manageable, 'strongly circumscribed and framed' (ibid. 63) and as camera-friendly and constructed, no doubt, as the filmic location. Its opposite, heterogeneous space 'with blurred boundaries' in which tourists mingle with locals (ibid. 64) implies, however, a liminality whose social, linguistic and other exchanges the paraphernalia of filming has greater difficulty in penetrating. By the same token, Edensor's cultural intermediaries, the local go-betweens, not dissimilar to Smith's (1978: 4) 'marginal men' who challenge the binary opposition of hosts and guests by suggesting a subtler and more complex interaction between tourist and toured, find little room for themselves in television reports. There, television reporters are the intermediaries, standing between locals and viewers and often obscuring the former, they themselves being linguistically and culturally visitors. True 'marginal men' are likely to liaise off-camera with production crews, acting as location 'fixers' or temporary members of the film unit. Thus the serendipitous and reciprocated glance between visitor and local which Chaney (2002) suggests as a preferable metaphor to the gaze to account for tourist behaviour is unlikely to figure in television's encounter with the destination, given too that the unblinking television camera lens rarely invites, let alone elicits, interaction from anyone other than the reporter who addresses it directly.

As a genre, therefore, holiday programmes have required themselves to adopt an essentially scopic approach which has privileged sightseeing and which does not necessarily reflect the more complex tourist discourses that current commentators identify. Their signifying conventions suggest a reliance on familiar, self-referential images of place and of genre, reflecting the collusion between tourists and travel photographers in the construction and reinforcement of place as touristic resource which Crawshaw and Urry (1997) have identified. Yet despite the importance of pictures, television, informed by discourses of radio and of journalism, has traditionally privileged words over pictures (Ellis 1992). Cronin's study of the problems of translation in travel writing suggests that most tourists gaze because they do not understand the language of the object of their gaze, arguing that 'the experience of travel in a country where the language is unknown to the traveller will be heavily informed by the visual. If you cannot speak, you can at least look. Sightseeing is the world with the sound turned off' (Cronin 2000: 82). Further, in any process of translation there remains an imbalance between the tourist and the toured since 'speakers of major languages are more likely to expect others to speak their language ... In other words, for powerful languages, the Other is always already translated' (ibid. 95). Television too uses its own language, verbal every bit as much as visual, to make such translations.

When it comes to sightseeing, television's rules are well established. The expected signifiers of place are shown in order to confirm the place. Paris has been signified (*Holiday*, 4 February 1992) by a shot of the Eiffel Tower and comedian Jimmy Mulville's words, 'of course no holiday film of Paris is complete without a long lingering shot of this thing. What's it called again?' In a report from Holland (*Wish You Were Here?* 13 February 1995), presenter Anna Walker announced herself 'unashamed' to insist on the appearance of clogs, tulips and bicycles on screen, and then offered a series of canal views of Amsterdam accompanied by the words, 'everywhere you look in Holland there's a shot which could be a Dutch Masterpiece'. Bern, with no well-known sights, had to be illustrated (*Wish You Were Here?*, 10 February 1992) by signifiers of general Swissness; trams, cake shops, towers with animated clocks, amplified by John Carter's almost apologetic words, 'a capital that doesn't overwhelm you'.

In the past this self-referential convention was occasionally subverted in C4's less mainstream, less tourist product driven, and now discontinued, *Travelog*. This offered the televisual equivalent of the travel essay and provided imaginative, and often quirky, responses to place. It did not feature holiday products *per se*, and that, given increasing budgetary constraints[2] from its commissioner C4, may have brought about its demise in 1997. At the end of a report on New Zealand (*Travelog*, 15 February 1995) Pete McCarthy alone in a bar addressed the camera. 'You know, we came to New Zealand to make a film without a single sheep in it, and it turned out to be surprisingly easy. Wherever you're travelling it's always worth making that extra effort to get beyond the cliché.' As he spoke two sheep wandered into the bar apparently unseen by him. As he finished they were heard baaing as he turned back to his drink. In general, however, the television camera makes do with more predictable images. It is, just like tourists themselves, a consumer of recognizable commodities.

Holiday programmes' discursive axes

Holiday programmes are heavily reliant on the scopic. Its opposite, the empathic, is accorded little space in a genre where short filming and short running times are the norm. Their reports have given meaning to the tourist destination conventionally through an underlying system of binary oppositions. The discursive axes of these oppositions operate in the microcosm of holiday programmes much as they have been perceived by the academy to operate in the macrocosm of holidaymaking. They are tourist/traveller, authentic/inauthentic, liminal/everyday, core/periphery.

The tourist/traveller axis articulated amongst others by Boorstin (1961) and Fussell (1980) has been reflected in the anti-tourism implicit in the genre's treatment of sightseeing. Seeing the canon of the sights of high culture, travelling 'off the beaten track' (Buzard 1993), was to be taken seriously. Fulfilling high-cultural obligations with the breviary of the guidebook, it was implied, was the penance to be paid for the perceived fall into tourism. Such anti-tourism was in keeping with the BBC's public service ideology of informing, educating and entertaining, an establishment ideology which Hall (1986) has suggested may have reflected popular culture but which made no commitment to it, and one which in turn, at least before deregulation of broadcasting, also informed ITV (Scannell 1990). There is, therefore, an ambiguity about sightseeing in a programme genre which claims to reflect a popular leisure activity. So, if it has to be done, it is done without unnecessary curiosity, a duty rather than a pleasure. This was highlighted by Sue Cook in Beijing (*Holiday* 7 January 1992) who suggested of the Forbidden City that 'you can allow a whole day ... and you still won't see it all'. The possibility of returning the next day was not mentioned. John Pitman's search (*Holiday* 21 February 1995) for anything to see on the island of Fuerteventura was finally rewarded when one isolated church was tracked down and specially opened so that he could see its magnificent gilded reredos. Gazing on it appeared to be sufficient. No information was offered about its age, style or significance. Its validity as a sight lay in its Spanishness.

The genre's authentic/inauthentic discursive axis reflects MacCannell's (1976) elaboration of Goffman's (1959) concept of front regions which are open to all, and back regions which are accessible only to insiders, in which he suggests that '[i]nsight, in the everyday, and in some ethnological sense of the term, is what is obtained from one of those peeks into a back region' (MacCannell 1976: 102). However much that binary opposition of inside/outside has been challenged as failing to recognize the fuzziness between producers and consumers, hosts and tourists (Boissevain 1996, cited in Coleman and Crang 2002: 5), the scripts of presenters still make much of their engagement with the 'real' life of the destination. Back regions can be opened up *to* the camera's gaze or indeed *by* the camera's gaze. The distinction is important. In presenting tourist destinations in the main as commodities, holiday programmes are reactive rather than proactive, more likely to observe what is on offer than to search for what is kept from, or does not seek, its gaze. Authenticity is, however, invoked by commentary often enough even if remaining undefined, and regardless of how staged the sequence in question may be. The very fact that it becomes a pro-filmic event, performed for and validated by the camera and articulating with its audience suggests that the theatrical metaphor of willing suspension of disbelief rather than the suggestion of duped consumer might well be applied to programme makers, viewers and tourists alike to define their relationship to such productions.

The behaviour of local inhabitants, who by definition inhabit or have access to the generality of a place's back regions, is often instanced in programmes as a marker of authenticity, although these locals remain for the most part either out of vision or in the background. Thus Julia Butt offered this piece of advice about Vienna (*Holiday* 28 January 1992). 'Do as the Viennese do and set off to see the sights on foot.' It is hard to evaluate such advice. Does her suggestion imply that the Viennese regularly sightsee in their own city, or that travel by foot is the favoured means of getting from place to place? *Summer Holiday* has offered 'insider' knowledge in the form of hints and tips to various destinations 'from the people who actually live there', while *Holiday, You Call the Shots* is based on viewers setting the agenda from their own experiences for what the presenters should visit, this apparent interactivity adding 'authenticity' to what are, in fact, a series of predictable and to all intents and purposes pre-structured reports from cities like Barcelona (25 September 2001) and Amsterdam (1 October 2001). These offered little more than an exhortation to visit, in the former, the Church of 'La Sagrada Familia' and dine at the much-touted restaurant 'Los Caracoles', and, in the latter, to visit the Anne Frank House and enjoy a canal cruise.

Holiday programmes do however offer their own occasional glimpses of back regions. Julia Mackenzie (*Wish You Were Here?* 23 January 1995), having already reported that 'Umbria is more authentic than Tuscany' without offering any amplification, introduced a sequence in a hotel with the hushed and reverential words, 'It's seven in the morning, and I've been allowed into the kitchen to watch the daily preparation of the pasta.' The sequence ended with her saying to the pasta maker, 'Thank you for sharing your secrets with us.' The fact that a cooking holiday was being featured, in which daily demonstrations, and thus an organized and regular sharing of secrets with people prepared to pay for the privilege, were on the menu, was not allowed to detract from this moment of staged backstage authenticity. Eating, since it is one experience common to tourists and locals, offers tele-vision regular opportunities for crossing from front to back regions, although such crossings are more likely than not to be staged for a camera whose intrusive presence may smack of negotiated access rather than serendipity, and are unlikely to feature extended encounters between tourist and toured.

The discursive axis of liminality articulates in holiday programmes as in holidaymaking with inversion of the everyday, with transgression and licence. Crouch (2002: 217) in positing tourism as an encounter with space offers a reminder that '[t]he world is grasped through the body and the world is mediated through the body'. Yet Shields (1991: 98) argues that even in marginal places social regulation 'tends to moderate the inversions and suspensions of the social order', while Stallybrass and White (1986) identify the bourgeois contradiction of disgust and desire implicit in carnival. Holiday programmes offer a very limited collusion with licence, and the transgressive excesses of resort club culture are more likely to be seen in late night 'fly on the wall' documentary series like ITV's *Ibiza Uncovered* or *Holiday Reps*. In one example of transgression, John Pitman (*Holiday,* 21 February 1995) was shown fully clothed on a naturist beach in Fuerteventura interviewing an unclothed holidaymaker. When talking, the naturist was shot head and shoulders only, but the interview was established by a full length shot of both taken from behind the interviewee and clearly, but discreetly, showing him to be naked. Television's often self-imposed codes of censorship thus ensured that the modesty of its audience was not outraged, and went even further in Pitman's reassurance that naturists kept far away from family beaches. The licences of liminality are modified before they can be shown to the viewers.

The final discursive axis of the genre is that of core and (pleasure) periphery, a touristic axis suggested, but insufficiently theoretized, by Turner and Ash (1975). True to the discourse of Orientalism (Said 1978) holiday programmes constantly seek to apply the values of home to give meaning to destinations abroad. Recognizing that the peripheral might not be 'civilized' enough, Anneka Rice introduced David Jessel's report from Malaysia (*Holiday* 10 March 92) with the words:

> 'Do you book a holiday to somewhere remote and exotic and then immediately wish that you'd gone for somewhere a little more familiar and mundane? Perhaps you long to wander somewhere far from the beaten track but at the same time crave the comforts of proper plumbing and shopping malls. Well David Jessel may have found the answer to all your problems ...'

Jessel picked up the theme with a postcolonial commentary which referred to the Padang and its cricket players in Kuala Lumpur as 'a safe little suburb in the heart of Asia' and encouraged viewers to eat out in the night markets which are no less safe than hotels, bidding them, 'have fun, be brave, eat out'.

The challenges of the Third World are greater. In Kerala Jill Dando (*Holiday* 2 January 1996) was shown walking through a run down alley with broken and missing steps with the words 'it's bad enough by day, but at night ...', and later reduced the state capital to a generality in a brief sentence illustrated by a fume-filled traffic jam of ancient looking lorries and bullock carts, 'Trivandrum is like any other Indian city ... chaotic.' Further up the coast in Goa, Paul Gogarty (*Holiday* 21 February 1995) visited a cashew nut factory 'that somehow seems to have skipped the industrial revolution'. In Salvador, Jill Dando reminded viewers (*Holiday* 9 January 1996) that Brazil is a Third World country and that crime is rife, making it advisable to stick to the tourist areas. There is a wary *schadenfreude* in much of this reporting, and the implicit suggestion that such visits are likely to make home seem all the nicer.

Presenters, show business and celebrity

Most destination reports feature a single presenter addressing the camera. Such a mode of address, using the 'close up' so suited to the domestic television set, constructs an intimate and informal relationship with the viewer (Corner 1996: 74). The presenter is, of course, a performer, part of what Postman (1982) has termed The Age of Show Business, in which '[o]ur priests and presidents, our surgeons and lawyers, our educators and newscasters need worry less about satisfying the demands of their discipline than the demands of good showmanship' (Postman *op.cit.*: 100).As with Boorstin's (1961) observations on the media's consumption of the pseudo-event, Postman reflects an American culture further down the line than yet holds in Britain perhaps, but there has been a significant blurring between show business and documentary in British television. 'Docu-soap' and 'reality television' both create and celebrate celebrity and involve viewers in a relationship which parallels 'the depth of the virtual relationship which viewers enjoy with [soap opera] characters (either positively or negatively), and the sense of coexistence between real and fictive worlds' (Corner 1999: 59).

In 1997 it was reported that from the following year Anthea Turner, former presenter of BBC1's *National Lottery Draw*, would replace Judith Chalmers on *Wish You Were Here?*. It has to be assumed that with her younger image the programme intended to attract new, and younger, male viewers. At the same time that Judith Chalmers (was) stepped down,[3] veteran travel journalist John Carter, aged like Chalmers sixty-one, and another *Wish You Were Here?* regular, was unambiguously fired. Paul Gogarty, the chief travel writer of the *Daily Telegraph* and himself a former, and fired, presenter of BBC's *Holiday* and therefore perhaps not an unbiased reporter, offered anecdotal evidence that *Wish You Were Here?*'s producers might have failed to read their audience's needs.

> The question is: do the majority watching these programmes want in-your-face youth pizzazz and anodyne 'infotainment' or something intelligent and authoritative? When John [Carter] was shopping last weekend a tanned 60-year-old came up to him and said: 'Don't they know that it's people like me who can afford these holidays, not 20-year-olds? And what I want to know is what a place is like, not how many changes of clothes some woman can get into a suitcase.'
>
> (Gogarty 1997: 13)

Television celebrity is grounded not in the auratic of the movie star but in familiarity and the ordinary, and Turner's status as a celebrity no doubt influenced her producers in their choice. Where once the authority of the presenter came from expert knowledge, now it comes increasingly from celebrity itself. The constructed domestic role, the appeal to their 'ordinariness', which news readers and chat show hosts, soap opera and comedy stars play in the 'flow' of broadcasting, is one whose function is to maintain the commercial interests of broadcasters and of the channel (Marshall 1997). Yet at the same time a celebrity lifestyle signifies distinction, status, wealth and aspiration (Rojek 2001). The celebrity may be as 'ordinary' as the viewer, but there is social distance in the encounter between the two, an encounter described by Horton and Wohl (1956) as 'para-social', that is a relationship of illusory intimacy constructed and exploited through the domestic role of the television set. It is also an encounter, as are holiday programmes, with consumption.

This billing appeared in *Radio Times* for the edition of *Wish You Were Here?* transmitted on 2 April 2001:

A special edition in which celebrities visit favourite holiday islands. Hollywood star Michael Douglas talks about his love of Bermuda, Bob Monkhouse reports from Barbados, *Coronation Street*'s William Roache takes a family holiday in Sardinia.

(*Radio Times* 2001: 91)

The programme itself was an uncharacteristic mismatch of styles, the Michael Douglas piece apparently bought from an American network as evidenced by its conversion from transatlantic NTSC to British PAL standard with consequent colour shift, and by its 'lifestyles of the rich and famous' signifying practices, while the Monkhouse segment was effectively a show business interview which happened to be set in the star's holiday villa. Roache's visit to the Forte Village Resort offered viewers little insight into the life of the island beyond the boundaries of the Village but did provide fans of *Coronation Street* with the sight of one of its stars bare-chested as he sunbathed, and the opportunity to aspire to a jet-set lifestyle as he and his wife debated the relative merits of 'The Dune' and 'Il Cavaliere' restaurants.

As soap operas increasingly dominate the schedules, and the battle for viewers, of BBC1 and ITV, both *Holiday* and *Wish You Were Here?* offer an increasing number of reports from soap opera stars, each naturally featuring those stars associated with their own channel. Rojek (2001: 23ff) has described the soap opera celebrity as a 'cele-actor', a fictional character who somehow embodies the *zeitgeist* but whose celebrity can create problems for the actor when public face threatens to stifle self. *Coronation Street* star John Savident's report from New Zealand (*Wish You Were Here?* 23 April 2001) reveals some of this tension. The opening shot showed Auckland's waterfront and skyline. As Savident walked into frame he declared, in theatrical received pronunciation far removed from his character, the Weatherfield butcher Fred Elliot's, northern accent, 'Auckland, city of sails', and, should his fans have failed to recognize the voice, he added, knowingly, 'Bit different from Weatherfield, isn't it?'. Almost immediately afterwards he was shown in company with his wife and her cousin as they prepared to visit the Sky Tower. 'Up that Kiwi phallic symbol; you must be joking,' with these words he mugged to camera and did a theatrical flounce out of shot. In Auckland Museum he adopted a hushed tone of solemnity while saying of the collection of Maori art that 'You can almost feel the spirituality [...] you can feel the atmosphere.' He did not, however, feel it necessary to spend any time looking at the artefacts or attempting to explain them to the viewers, and appeared happier in the next sequence, a wine tasting, when as he wandered through vineyards, glass in hand, he parodied himself as an old thespian smacking his lips theatrically and emoting the words 'oh my dear boy, oh, oh,' as he tasted the wine. For a celebrity even when on holiday, performance is perhaps more important than informed response to place.

Conclusion

When the whole world has been visited, its images endlessly recirculated, what remains? MacCannell (1999: 191) suggests that we have all become tour guides when we introduce others to something of our world, and television's increasing reliance on non-expert reporters suggests that it is the variety of (often celebrity) experience that is the new televisual tourism. If the tourist destination has been progressively exhausted by the cycle of self-referentiality of the television camera and has exhausted those behind that camera, then its transformation into a *locus* for gazing on the lifestyles of celebrities offers new programme-making opportunities. Show business and supposedly factual reportage have had to coexist increasingly in the last ten years, not only in holiday programmes but also across television's broader spectrum. Wells (2001) draws attention to the fact that the appointment to head of a newly created BBC department of 'factual entertainment' of former *Big Brother* producer Conrad Green created such negative publicity that the job had had to be renamed 'head of entertainment development'. The message is clear, however, that show business and celebrity will continue to inform factual programming for the immediate future. Holiday programmes are not of particular depth, and yet Gogarty's (1997) words which began this chapter, and from which it takes its title, do have a resonance. There does appear to be a 'dumbing down', a diminishing of the amount of time given over to explaining a destination and an increase in the time given over to celebrities whose sole reason for being on screen is that they are celebrities, part of a solipsistic world where place becomes a space for performance.

Let BBC1's *Holiday, You Call the Shots* from Barcelona (25 Septembeer 2001) have the last word. The conceit of the programme is that its three presenters, science and lifestyle programmes presenter Kate Humble, youth and music presenter Jamie Theakston and comedian Rowland Rivron are required to go wherever viewers, drawing on their own holiday memories, suggest. Early in this show Humble is seen in 'close up' at a café table, mobile phone to her ear, Barcelona and its inhabitants in soft focus in the background. She claims to be in touch with former FC Barcelona manager Bobby Robson, and after chatting about his memories of Barcelona, largely about the weather, she asks Robson if he can arrange for Theakston to visit the FC Barcelona's Ground. Moments later Theakston has joined Humble at the table. She says, 'Bobby Robson, I spoke to him today.' 'You're joking,' replies Theakston, feigning surprise for the camera. Humble explains her ploy. Theakston counters with, 'I've got something for you. I've just got [*sic*] a call from Tamzin Outhwaite [of BBC1's *EastEnders*]. She knows a swimwear shop which serves frozen Margaritas while you shop. I thought that would be just up your *Strasse*.' With that, busy presenters that they are, they down their drinks, gather their mobile phones and rush off for another encounter with, and celebration of, celebrity. Today is Tuesday 25th so it must be Barcelona. Of course it must. It says so in *Radio Times*.

Notes

1. Source: BARB [Broadcasters' Audience Research Board] 1992–2002 *et passim*.
2. *Travelog*'s editor Jenny Mallinson Duff drew the writer's attention to the beginning of budget cuts in correspondence in 1992.
3. The *Guardian* (11 April 1997) was required to print an apology for having suggested (8 April 1997) that she had been fired, and to make clear that she would remain a regular presenter of reports.

References

Batten, Charles L. (1978) *Pleasurable Instruction*, Berkeley, CA: University of California Press.
Boissevain, Jeremy (ed.) (1996) *Coping with Tourists: European Reactions to Mass Tourism*, Oxford: Berghahn.
Boorstin, Daniel J. (1961) *The Image*, London: Weidenfeld and Nicolson.
Buzard, James (1993) *The Beaten Track: European Tourism, Literature, and the Ways to 'Culture'*, Oxford: Clarendon Press.
Chaney, David (2002) 'The Power of Metaphors in Tourist Theory', in Simon Coleman and Mike Crang (eds) *Tourism, Between Place and Performance*, New York: Berghahn Books: 193–206.
Clifford, James (1992) 'Travelling Cultures'. In Larry Grossberg, Carry Nelson and Paula Treicher (eds) *Cultural Studies*, New York: Routledge, pp. 96–118.
Coleman, Simon and Crang, Mike (2002) 'Grounded Tourists, Travelling Theory', in Simon Coleman and Mike Crang (eds) *Tourism, Between Place and Performance*, New York: Berghahn Books, pp. 1–17.
Corner, John (1996) *The Art of Record*, Manchester, Manchester University Press.
——(1999) *Critical Ideas in Television Studies*, Oxford: Clarendon Press.
Crang, Mike (1999) 'Knowing, Tourism and Practices of Vision', in David Crouch (ed.) *Leisure/Tourism Geographies: Practices and Geographical Knowledge*, London: Routledge: 238–256.
Crawshaw, Carol and Urry, John (1997) 'Tourism and the Photographic Eye', in Chris Rojek and John Urry (eds) *Touring Cultures: Transformations of Travel and Theory*, London: Routledge, pp. 176–195.
Cronin, Michael (2000) *Across the Lines: Travel, Language, Translation*. Cork: Cork University Press.
Crouch, David (2002) 'Surrounded by Place: Embodies Encounters', in Simon Coleman and Mike Crang (eds) *Tourism, Between Place and Performance*, New York: Berghahn Books, pp. 207–118.
Culler, Jonathan (1981) *In Pursuit of Signs*, London: Routledge.

Edensor, Tim (2001) 'Performing Tourism, Staging Tourism: (Re)producing Tourist Space and Practice', *Tourist Studies* 1(1): 59–81.

Ellis, John (1992) *Visible Fictions*, London: Routledge.

——(2000) *Seeing Things: Television in the Age of Uncertainty*, London: I.B.Tauris.

Franklin, Adrian (2001) '*The Tourist Gaze* and Beyond: An Interview with John Urry', *Tourist Studies* 1 (2): 115–131.

Franklin, Adrian and Crang, Mike (2001) 'The Trouble with Tourism and Travel Theory?', *Tourist Studies* 1 (1) 5–22.

Fussell, Paul (1980 *Abroad: British Literary Travel Between the Wars*, New York: Oxford University Press.

Goffman, Erving (1959) *Presentation of Self in Everyday Life*, New York: Doubleday.

Gogarty, Paul (1997) 'Have Youth, Will Travel', *Daily Telegraph*, 19 April 1997: *Weekend,* 13.

Hall, Stuart (1986) 'Popular Culture and the State', in Tony Bennett, Colin Mercer, and Janet Woollacott (eds) *Popular Culture and Social Relations.* Milton Keynes: Open University Press, pp. 22–249.

Horton, Donald and Wohl, Richard (1956) 'Mass Communication as Para-Social Interaction: Observations on Intimacy at a Distance', *Psychiatry*, 19: 215–29. Reprinted in J.Corner and J.Hawthorn (eds) (1993) *Communication Studies* (4th edn) London: Arnold, pp. 156–64.

MacCannell, Dean (1976) *The Tourist: A New Theory of the Leisure Class*, New York: Schocken Books.

——(1999) *The Tourist: A New Theory of the Leisure Class*, with a new Foreword by Lucy L Lippard and a new Epilogue by the author, Berkeley and Los Angeles, CA: University of California Press.

——(2001) 'Tourist Agency', *Tourist Studies* 1 (1): 23–37.

Marshall, P. David (1997) *Celebrity and Power: Fame in Contemporary Culture*, Minneapolis, MN: University of Minnesota Press.

Osborne, Peter (2000) *Travelling Light: Photography, Travel and Visual Culture*, Manchester: Manchester University Press.

Postman, Neil (1982) *The Disappearance of Childhood*, New York: Delacorte Press.

Morgan, Nigel J. and Pritchard, Annette (1998) *Tourism Promotion and Power: Creating Images, Stealing Identities*, Chichester: John Wiley.

Radio Times (2001) Week Commencing 31 March 2001.

Rojek, Chris (2000) *Leisure and Culture.* London: Macmillan.

—— (2001) *Celebrity*, London: Reaktion Books.

Said, Edward (1978) *Orientalism*, London: Routledge.

Scannell, Paddy (1990) 'Public Service Broadcasting: the History of a Concept'. In Andrew Goodwin and Gary Whannel (eds) *Understanding Television*, London: Routledge, pp. 11–29.

Seaton, A. V. (1989) Freebies? Puffs? Vade Mecums? or Belles Lettres? The Occupational Influences and Ideologies of Travel Page Editors, Newcastle: Centre for Travel and Tourism, in association with Business Education Publishers.

Shields, Rob (1991) *Places on the Margin*, London: Routledge.

Smith, Valene (1978) 'Introduction' in Valene Smith (ed.) *Hosts and Guests: The Anthropology of Tourism*, Oxford: Blackwell, pp.1–14.

Stallybrass, Peter and White, Alan (1986) *The Politics and Poetics of Transgression*, London: Methuen.

Sweeting, Adam (1997) 'Last Night's TV: Fly me to Freebie', *Guardian* 23 April 1997, *G2T*, 12.

Turner, Louis, and Ash, John (1975) *The Golden Hordes*, London: Constable.

Urry, John (1990) *The Tourist Gaze*, London: Sage.

——(1995) *Consuming Places*, London: Routledge.

——(2000) Sociology Beyond Societies: Mobilities for the Twenty First Century, London: Routledge.

——(2002) *The Tourist Gaze* (2nd edn), London: Sage.

Wells, Matt (2001) 'There's no such thing as reality TV', *Guardian* 5 November 2001, *G2 Media*, 2–3.

13 Tourists and television viewers

Some similarities

Solange Davin

The words 'television viewer' bring to mind caricatures of lazy couch pota-toes, glued to their seats for hours on end, trying to escape reality by passively staring at addictive and mindless drivel on the screen. The word 'tourist', on the other hand, evokes visions of dynamic doers eager to explore new horizons and to partake in exciting, life-enhancing experiences. Yet viewers and tourists have more in common than these popular stereotypes may lead to believe. This chapter overviews some shared ground between the two, drawing on the existing literature and on a reception study of the American medical drama *ER*.[1]

Backstage spectacles

The hallmark of television and of tourism[2] is, first and foremost, spectacle. As leisure industries and consumption continue to grow, so does the need for spectacle. This rising demand has led to the marketing as tourists spots of areas which had not previously figured on – in some cases which had been carefully kept away from – tourists' lists of 'things to see' or publicity brochures, like sewers (in Paris for instance) which are now being promoted as appealing venues and which welcome thousands of visitors.[3] This is paralleled by a public upsurge of curiosity about hitherto invisible territories, mundane everyday spaces, 'back regions', to use Goffman's (1974) expression. The exploration (for example) of formerly hidden aspects of work places is a staple of both organized tours (fire stations, farms, coalmines, factories, etc. (see Rojek 1993: 152)) and of 'reality programmes' such as docusoaps, many of which aim at showing the ins and outs of the working life of various groups of 'ordinary people' (on ships, in airports, in offices, in hotels etc.).

Insights into the backstage of medicine and of hospitals can be gleaned from medical dramas such as *ER*. Far from offering pictures of god-like, perfect physicians, of all-knowing and all-powerful heroes – Goffman's 'front regions' – as did some early medical dramas, *ER* takes a 'warts and all' approach to medicine and to doctors, who are depicted as prone to making mistakes, as stressed by personal problems, as riddled with medical and ethical dilemmas, as anti-heroes, as flawed beings – in short, as human:

Doctors and nurses are human, for example Carter's ambition and naivety, Greene's marital problems…
The doctor here is human while he is usually presented in the style of Dr. Welby, a head above patients, with grey hair, full of wisdom, with a spotlessly clean car without a single cigarette end. Here you have one doctor who drinks, who has not shaved, Doug Ross, who has plenty of charm but lots of personal worries but who is wonderful with children while he obviously does not know how to deal with adults…

Moreover, being located in one of a small number of American public hospitals catering for the most impoverished section of the population, *ER* is ideally placed to reveal the flip-side, the backstage, of the American Dream propagated by Hollywood-style serials like *Dallas* or *Dynasty*. Viewers praise the non-PC dimension of *ER* which imparts glimpses of the climate of poverty, of exclusion, of violence encountered by the many homeless and unemployed in North America:

We keep seeing people who need social workers, who cannot afford healthcare, children's bodies full of gun shots, abused children. It seems to me that ER gives a rich and realistic portrait of American society.

Television and tourism as positive/negative forces

These 'politically incorrect' images can be found in other recent programmes (see Winckler 2002), notably in some police series where officers are portrayed as less than honest and competent. This propensity for displaying the non-PC in and beyond entertainment gives television, it has been argued, a progressive dimension. Television is liable to reappraise and to challenge the role and the functioning of institutions, be it the police or the medical or legal systems, to expose deceptions, mistakes, double-standards that officials and corporations would prefer to see forgotten, to dispute the motivations and the *bien-fondé* of policies, to demand – and to obtain – public investigations into controversial events or decisions. Television has forced governements to admit to cover-ups (see for example Reilley and Miller 1997 on BSE) and, on occasion, it has even contributed to their downfall – Watergate springs to mind.[4] Equally, museums can challenge taken-for-granted versions of events, thus courting controversy and promoting debate, as in the case of the 'Enola Gay affair' at the Smithsonian Institution in Washington, where two antagonistic discourses concerning the end of the Second World War were found to clash (Zolberg 1996).[5] Conversely, both tourism and television have been accused of being conservative agents supporting the status quo and dominant worldviews. In this model, viewers and tourists alike are deemed to be manipulated by propaganda, deceived by false transparency, fooled into believing fanciful narratives (for example, of a benign bio-medicine, of an immutable past), confused by contradictory messages, their attention diverted from serious political matters by popular, worthless (mostly American-style) mass entertainment symptomatic of and fostering a general 'dumbing down' via 'wall-to-wall Dallas' and 'Disneymania'.

Television and tourism have been praised for their positive features. The latter, it has been said, preserves wildlife and the countryside, strengthens local lifestyles, promotes cross-cultural understanding and tolerance, helps vulnerable groups etc. Television defuses violence, provides education for the many, improves the public's knowledge of the world and of themselves, offers good role models for adolescents, enhances children's cognitive skills, and so on. But both have also been blamed for an array of destructive deeds (often the very reverse of the previous): ruining wildlife and the countryside, destroying ancient traditions, exploiting the underprivileged, fuelling cultural misunderstanding, causing copycat violence, providing negative role models, preventing children from engaging in activities deemed to be more valuable such as reading.

Gendered leisure?

In the Sociology of Tourism as in Media/Television Studies dubious gender differences have been postulated, and found wanting. As far as television is concerned, it has long been assumed that women's preference goes to entertainment, to fiction, particularly to the melodramatic trivia of soap opera, while men are more inclined to watch factual broadcasts such as news bulletins and documentaries. Similarly, some forms of tourism, like ecotourism, have been regarded as a primarily masculine pastime.[6] Yet research has undermined these claims: many women take part in ecotourism (Fennell 1999: 58–59), watch news bulletins (e.g. Alasuutari 1992: 567) and other allegedly 'masculine' programmes like sports (Poynton and Hartley 1993) or science fiction (Jenkins 1992). Equally, numerous men follow cooking programmes (Meyrowitz 1986: 88) and soap operas: 20 to 30 per cent of soap viewers in the US are male (Harrington and Bielby 1995: 15) and over 40 per cent in Brazil (Kottak 1990: 25).[7]

The pedagogy of mass entertainment

The tourism and television industries share some pedagogic strategies. Exhibitions such as 'living museums' seek to bring history to life (as often as not with the help of actors). Visitors are encouraged to enjoy themselves and, at the same time, to learn by participating in the past and by absorbing its atmosphere (see Rodaway 1995: 256–7, Rojek 1993: chap. 4). This echoes the 'infotainment' trend of genre *mélanges* and audiences' taste for hybrid broadcasts, for programmes which are at once entertaining and educative (see for instance Elkamel 1995; Bouman, Mass and Kok 1998). Thus many informants enjoy the recreational side of *ER* while gathering medical, health promotion, socio-medical and social data from its informative elements, as these quotes indicate:

I watch ER for its entertainment value, and it gives 100 per cent. But I also enjoy it because I am interested in medicine and in patients' views and feelings.

The main interest of the series (apart from entertainment and stress) is that through the emergencies we discover the social life in the US, the problems of insurance, of violence, of abuse.[8]

In addition, tourists as well as viewers report extracting knowledge through 'emotional' tactics such as identification. Just as a visitor commented on leaving the Jorvik Experience centre in York: 'I felt I was a Viking mother back there. It was ever so hard for us, back then' (Rodaway 1995: 258), *ER* followers identify with characters, especially with medical students with whom they feel a strong affinity, a complicity, due to a (perceived) common status of newcomer on the ward:

Spectators were thrown in the whirlwind of situations alongside the students. They could be the spectator who has crept inside the emergency department. Their beginners' eyes are ours.
The student is the beginner prototype, to whom everything happens, who is ignored or patronized, but who evolves. I really like him. He resembles the spectator.

As the Jorvik tourist, by 'participating in' the (vicarious) experience of 'being a Viking mother', discovered how difficult life was 'for us back then', through the eyes of the young medical trainees, by sharing their fears and their hopes, their errors and their frustration, their achievements and their progress, their deception when they make a mistake and their joy when they succeed, *ER* fans gradually discover how a hitherto mysterious casualty ward functions.

'Hyper-realism' in television and in tourism

Both tourists and viewers may take prior mediated experiences as a reference point.[9] Tourists 'test' attractions against previously visited attractions. For example, instead of comparing Euro Disney features to the fairy tales, historical narratives or real places on which they were based, visitors contrast them to similar thematic representations at Disney World (Rodaway 1995: 255–6). In a similar fashion, other broadcasts may become a benchmark for *ER*:[10]

Before the Atlanta Olympics, there was a documentary on the US and everything looked like ER.
I have never been to the States but I know from other series.

Counter-intuitive reversals also occur whereby a factual broadcast, or even real life itself, is compared to the fictional show which thereby becomes the reference point:

If I did not know that the Real ER is a documentary I would think that it is another ER. The characters look like ER characters. At the beginning, it's Dr. Greene, and the nurse could be Carol.
People say to my wife, who is a nurse, why don't you wear a green jacket as they do on television?

Paradoxically, it seems that the mediation of experience gives it an extra – a more real – quality: for Woolley (1993: 195), who witnessed the immediate aftermath of the Clapham rail disaster, it was nevertheless on television that 'the event happened', his own experience having a 'lower meaning'. Similarly, a police officer changed his testimony in the 'Rodney King trial'[11] after watching the videotape of King being attacked by the police, which he too thought had a sense of higher truth than his own *vécu* (Fiske 1995: 130). Mediated reality has become so omnipresent that some tourist sites provide their guests with a mediated version of their attractions in parallel to the real ones. Thus tourists at the Niagara Falls can admire the Falls either 'first-hand or vicariously through the three-dimensional film at the Imax theatre' (Hannigan 1995: 193). Tourists may soon come to find reality disappointing, warns Rojek (1997: 54), while Eco (1985: 46–7) fears that travellers on 'nature' excursions may miss the wildlife on display at Disneyland. Tourism may then be accused, as the small screen has often been, of fostering a mood of escapism.

To fulfil tourists' expectations that reality should match its representations, many social practices – songs, dances, markets, festivals, celebrations – have been re-created, modified, developed, accommodated (and perhaps simply invented?), many artefacts large and small – from jewellery and clothing to boats and houses and even entire villages – have been redecorated, embellished, transformed, so that visitors can witness and 'absorb the atmosphere' of the unique, 'authentic' culture which they have come to consume (see Boissevain 1996) (although they are clearly aware of these (re)construction processes, as will be seen below). The same applies to television. Unlike the era of 'paleo-television' (Eco 1985) whose characteristic was to present a direct, pure, unconstructed reproduction of the world, the 'neo-television' of the late twentieth and twenty-first centuries actively participates in its construction:

> To be sure ... a child fell in a well, and *it is true* that he died. But everything which happened between the beginning of the accident and the death so happened because television was here. The event, televisually caught at birth, became a *mise-en-scène*.... The presence of cameras influences the course of events.... We could even wonder what would have happened if television had not been there...
>
> (Eco 1985: 150–152)

Furthermore, reality may be intentionally reorganized for maximum screen appeal, from sporting events (see Morse 1983) to court cases like OJ Simpson's where barristers turned media-stars addressed their arguments to the media as well as – perhaps more than? – to the members of the jury (see Sherwin 2000) through 'media-events'[12] like the wedding of Prince Charles and Diana Spencer, where:

> it was absolutely clear that everything which was happening, from Buckingham Palace to St Paul's Cathedral, was planned for television. Interpretation, manipulation, preparation for television preceded the activity of the cameras. The event was born fundamentally false, ready to be filmed. The whole of London was prepared as a studio, it was reconstructed for television [...] Increasingly, the natural event is approached in terms of transmission. The wedding of the United Kingdom Prince Regent proves this hypothesis.
>
> (Eco 1985: 152–154).

With the rise of heterogenous programmes like docusoaps where the camera re/constitutes incidents which had no witnesses or incidents which 'only almost happened', thus generating 'media-truths',[13] with the proliferation of 'reality-television' and of a myriad of cryptic genres (false documentary, mockumentary, news fiction, faction, soapumentary, dramedy, pseudo-documentary, infomercial, quasi-documentary, fictional documentary, fictionalized documentary, docutainment etc.), with several 'faked documentaries' scandals in the late 1990s (see Maddox 1999), with fictional events making news headlines (for example the 'Who shot JR?' plot in *Dallas*), with politicians taking part in soap operas (ex-President Ford and Henry Kissinger in Dynasty) or becoming entertainment celebrities (for instance Jerry Springer) while actors turn politicians (Ronald Reagan is probably the most notorious) and fictional characters play a key role in elections (as in the (in)famous case of the 'argument' between the US Vice-President Dan Quayle and Murphy Brown, a sitcom character) (see Fiske 1995), the 'real' and the 'mediated', fact and fiction, document and spectacle, irrevocably merge and implode. It is the 'dissolution of TV into life, of life into TV [...] a sort of fantastic telescoping, of collapse of one into the other of two traditional poles... implosion' (Baudrillard 1981: 54–5).[14]

Boundaries between tourism and everyday life are getting blurred in many ways. First, just as tourists awaiting to reach the next destination of their journey may be suspended in the unknown, in the unstable 'new world order' where long-standing customs and institutions – the nuclear family, life-span employment, unions etc. – have been eroded, familiar routines could collapse overnight and a permanent question mark hangs over tomorrow. Second, leisure is increasingly overlapping with toil: thus 'holiday packages' may include periods of work as what used to be known as 'voluntary work' is re-labelled, for instance, 'conservation holidays'. Third, wherever they go, tourists encounter almost undistinguishable situations and procedures and cannot help having repetitive experiences (see Rojek 1993: 199–203): identical wide-bodied jets, airports, customs, transfer operations, high rise air-conditioned hotels, chain restaurants and stores, beaches, souvenir shops, shopping malls with their analogous fast food outlets and reconstructions of 'typically foreign villages'. The line between *Home and Away*, between work and rest, between boredom and excitement, between novelty and routine, between duty and freedom, between safety and danger, between the exotic and the ordinary, between the expected and the unusual, is gradually fading. As Urry (1995: 148) has argued, 'people are tourists most of the time whether they are literally mobile or only experience simulated mobility through the incredible fluidity of multiple signs and electronic images', signs which form a system 'in which images and stereotypes from the past and the future, from the locale and the globe, are implacably intermingled' and which make everyone 'a permanent *émigré* from the present' (Rojek 1993: 168). Postmodern individuals are thus constant (space and time) tourists.

In a 'hyper-real' world, beyond what MacCannell (1973) has called 'staged authenticity', tourism is enmeshed with reality in a more disturbing fashion. In a later book, MacCannell (1992: chap. 8) describes how an entire Californian community, Locke, was acquired by a multinational conglomerate and marketed as 'the only intact rural Chinatown in the US'. Its unfortunate residents, reluctant exhibits in their own hometown, were henceforth condemned to live under, to be probed by, the unremitting gaze of visitors. But is their plight so different from that of the citizens of Irvine, a corporation-owned Orange County 'hypersuburbia' where people also live under the surveillance of relentlessly inquisitive eyes, be it of local committees or of probing neighbours, and where everything, down to the smallest details – which breed of dog residents are (not) allowed to own, which shape their lawns may (not) take, which plants they may (not) grow, which colour they may (not) paint their houses etc. – is 'controlled by the strictly enforced rules of the planned community' (MacCannell 1992: chap. 2)?[15] The occupants of this panoptic 'demonstration city of the world' are as much on show in an 'unreal', artificial world, as much part of a 'living historic preservation', as restricted and as 'unfree' (ibid. 174), as the citizens of Locke. Residents in both communities have been 'trapped behind the corporate curtain' (ibid. 81), transformed into objects to be scrutinized, into a spectacle, into commodities. Moreover, with the omnipresence of CCTVs, it could be argued that we are all permanently being monitored, that we are all on continuous display. This trend is, of course, mirrored by the rising number of *Big Brother*-type programmes on the small screen where cameras (allegedly) capture, register and broadcast every minute, every minutiae, of the participants' lives.

Post-modern viewers and tourists

One should not conclude from the previous pages that tourists or television viewers fit the perennial stereotypes of naive sponges absorbing all images unthinkingly, naively confusing fact and fiction. What Macdonald (1995, see also for example Falk and Dierking 1998) found for museums, that visitors do not simply accept artefacts as they are exhibited, do not decipher them 'passively' but re-frame and interpret them in a manner which had not necessarily been anticipated by the designers, also applies to television. Telespectators are media-literate, astute, insightful. They produce subtle, elaborate interpretations.[16] Medical dramas, for example, are 're-genred' and re-read as mystery stories, as quizzes (viewers play at guessing diagnoses and treatments as soon as the symptoms/clues are revealed), as comedies (the style and content of 1950s and 1960s broadcasts now appear so quaint and old-fashioned that they are amusing), and as documentaries (from which, as already indicated, they gather a great deal of information):

I find it a very realistic series, that is, in the sense of 'documentary', which is very rare and interesting.
When I watch ER it's more like watching a documentary.

Postmodern tourists and viewers are fully cognizant of the practices of production and evaluate the spectacles which are presented to them in the light of this not inconsiderable knowledge. Thus, just as visitors of 'heritage experiences' were found to accept the authenticity of a 'fantasy of the past' in the full awareness that 'their details are "not really true"', that what they are seeing is a 'cleaned up' version of the world (Rodaway 1995: 255–256), *ER* viewers are critical of some sections of the scenarios which they deem exaggerated and unrealistic (notably the incredible number of heavy cases which arrive on the ward during the 50 minutes of each episode and the break-neck speed at which these patients are assessed and treated) but nevertheless remain adamant that the serial is realistic.[17] They know that these implausible elements are not a result of deliberate deception, of mistakes or of sloppiness but that they have been deliberately included in the narratives to heighten the level of dramatic tension, and that such excesses are a fundamental requirement of the genre. This explains, excuses, redeems them:

A series needs drama hence everything happens in the same place.
Although it is rare for a casualty to have so many disasters, the fiction requires it!

Conclusion

In the postmodern era it is impossible not to be exposed, directly or indirectly, to television, whether one owns a set or not.[18] Equally, we are all perforce permanent tourists of a kind. Postmodern viewers and tourists are

media/tourism-literate, sophisticated, ludic, ironic, shrewd, *both/and* creatures: both detached and involved, relaxed and demanding, distracted and
focused, distant and close, trusting and sceptical, bored and excited,[19]
learning and being entertained. As the spread of the entertainment industry
continues unabated across the globe, as entire regions are being turned into
giant theme parks, as the world becomes a complex web of intertexts and
hypertexts, reality, media and tourism are more and more closely intertwined. Tourists are viewers, viewers are tourists.

Tourism and television-watching may soon be almost undifferentiated.
One may stay at home and watch the soap opera *Coronation Street* on television or one may take a trip to Manchester to visit the *Coronation Street*
studios. One can practise 'armchair tourism' by watching the numerous
travel programmes on television or 'virtual' tourism on the Internet (see
Poster 1996: 30) or, better still, one can engage in the ultimate postmodern
expedition, 'collage tourism', a sort of 'mind-voyaging' whereby 'we can
manipulate the meaning of sites by dragging on different file indexes and
combining elements to create meanings which differ from broadcasts reports
of the sensation [...] an activity belonging to the ephemera of televisual
mass culture' (Rojek 1997: 63).

There is one last – but not least – similarity between telespectators and
tourists: little is known about them. A number of disciplines (Psychology,
Sociology, Anthropology, Economics, Media Studies, Cultural Studies) have
began exploring these phenomena, but empirical studies remain scarce.[20] Yet
existing findings clearly indicate that interpretation processes, be it of television broadcasts or of tourist sites, are complex and multi-layered, and that,
while the meanings which readers attribute to these spectacles may be unpredictable, sometimes idiosyncratic, often contradictory, they are always
intricate and dynamic. More in-depth research is needed if the substantial
savoir-faire of these voracious image consumers is to be thoroughly investigated.

Notes

1 This article does not seek to provide a comprehensive overview of tourists' or
   viewers' typologies, motivations or activities but, more modestly, to highlight
   some points of convergence between the two. The study is based on almost 200
   letters received in response to a message in television magazines asking viewers to
   explain why they watch ER (further details are available in Davin, forthcoming,
   2003, 2004, 2004a). Italics are verbatim quotes from informants.
2 Tourism has no simple, fixed meaning. There are many forms of tourism (see for
   example Smith 1989). Equally, tourists come in many shapes and forms (see
   Rojek 1993: chap. 5).
3. Some currently popular tourist spots were once deemed unfit for travellers such
   as the Lake District, which was regarded as 'the very embodiment of inhos-
   pitability' (Urry 1995: 193).
4. Watergate is not an isolated instance. In Brazil, demonstrations inspired by a
   telenovela led to the downfall of a corrupt president (Tufte 2000) and in

Queensland a documentary on corruption led to the collapse of the government (Stewart, Lavelle and Kowaltzke 2001). On resistance see, amongst others, Wayne (ed.) (1998), Dowmunt (ed.) (1993) for television; Boissevain (1996), Crain (1992, 1996) for tourism.

5. For details of the controversy see Bird and Lifschultz (eds) (1998: part IV).
6. Tourism was once deemed to be a male prerogative altogether (see Beezer 1993).
7. Three reasons can be suggested for these alleged gender differences. First, research subjects in Media Studies are often recruited in 'women's magazines' (e.g. Ang 1991) which is unlikely to encourage men to participate. Second, men may deny watching so-called women's programmes in the presence, and under the pressure, of peers (see Buckingham 1987: 197) (they are more likely to report it in questionnaires, see Tulloch and Moran (1986: 251)). Third, differences attributed to gender may be related to other factors (such as work patterns or lifestyle) (see Hoijer 1999: 183).
8. Viewers collect information about medical details – symptoms, diseases, treatments, drugs, about emergency procedures, health promotion messages, socio-medical issues (the American privatized healthcare system is compared to European Welfare State provisions with which viewers are familiar and found inadequate) and, as already seen, about social life in the United States.
9. This is not unusual: some of Philo's (1996) informants trusted mediated information on mental illness over their own experience. In the Cosby Show, Clare Huxtable as a lawyer and the Cosby family as 'middle-class' were seen as more representative than those met by viewers in real life (Press 1991: 110–1).
10. The same phenomenon occurred in my preceding Casualty research (see Davin 1999): *I compare Casualty to reconstruction shows. I watch 999 and Crimewatch and there you get the real attitude of the emergency services. They are authentic programmes.*
11. In March 1991, a black car driver, Rodney King, was stopped and beaten up by (white) Los Angeles police officers. The incident was filmed by a local resident on a home video camera. Later that month the officers were indicted on charges of excessive violence and in April 1992 they were found not guilty by a jury. The uprisings took place in Los Angeles. In August 1992, a grand jury indicted the officers of civil right charges and two were found guilty in April 1993 (see Fiske 1995: 14–18).
12. 'Media-events' are televised occasions (ceremonies like coronations, royal or celebrities' weddings, historic events like the moon landings, the shooting of John Kennedy, the 2001 terrorist attacks in New York, international competitions such as the Olympic Games, and so on) which transfix the world and which involve festive viewing and liminality (Dayan and Katz 1992). They are not to be confused with Fiske's (1995) notion of 'media events' (events whose reality lay in part in their mediation).
13. A typical example of 'media-truths' is the photographs of OJ Simpson in his escape car 'pointing a gun to his head and speaking into his car phone, an image that only a computer could produce for no camera could have been present to take it' (Fiske 1995: xxiii). Similarly reconstructed sequences in docusoaps, from the insomnia of a driving-licence candidate to the planned burglary which was accidentally prevented by the production team and subsequently (hypothetically) reconstructed (these programmes were discussed by viewers in On Air, BBC2, 7 December 1998, at the height of the docusoap fashion), have led to a barrage of criticism. Polemics continue to rage as to whether these are 'corrupted factual programmes' or whether these practices are innocent examples of 'ordinary media grammar'.
14. Fjellman (1992: 255–7), writing about Disney World, proposes a fourfold model of (Disney) reality (real real, real fake, fake real, fake fake). See Kepplinger and Habermeier (1995) for a 'genuine/mediated/staged' reality continuum.

15. One difference, of course, is that people choose to live in Irvine in (presumably) full awareness of these practices. Not everybody, however, is welcome in Irvine where the problem is more one of exclusion since 'there is no housing for the poor' (MacCannell 1992: 81).

16. For other examples of sophisticated viewers see for instance Buckingham (1997), Hill (1999), Just *et al.* (1996), Turnock (2000).

17. Almost all informants mention the subject of realism and there is consensus that *ER* is a highly realistic medical drama. But their concepts of realism go well beyond any direct screen–reality comparison (one example mentioned in the main text is how form and content interact when viewers assess *ER* realism – i.e. unrealistic excesses are discounted if deemed necessary to the genre) (see Davin forthcoming 2004, 2004a).

18. For a study of the small proportion of people who do not own a television set, see Castro-Thomasset (2000).

19 Some ER fans appear to be simultaneously fascinated and bored: 'You are taken by the suspense, will they make it, will they save the baby? It is always somewhat the same: we know what it's about, what to expect.' 'I did not want to watch but now I am spellbound. After a while all the episodes are alike and you get bored.'

20. There has been a surge in television reception studies in the past two decades, since Morley's (1980) seminal study of the 1970s current affairs programme *Nationwide*, but they nevertheless remain a minor (and underrated) part of the discipline.

References

Alasuutari, P (1992) 'I'm ashamed to admit it but I watch Dallas', *Media Culture and Society* 14 (4): 561–82.

Ang, I (1991) *Watching Dallas*, London: Routledge.

Baudrillard, J (1981) *Simulacres et simulations*, Paris: Éditions Galilée.

Beezer, A (1993) 'Women and "Adventure Travel" Tourism', *New Formations* 21: 119–130.

Bird, K and Lifschultz, L. (eds) (1998) *Hiroshima's Shadow*, Stony Creek, CT: The Pamphleteer's Press.

Boissevain, J (1996) 'Introduction'. In J. Boissevain (ed.) *Coping with Tourism*, Oxford: Berghahn.

Bouman, M., Maas, L. and Kok, G. (1998) 'Health Education in Television Entertainment', *Health Education Research* 13 (4): 503–518.

Buckingham, D. (1987) *Public Secrets*, London: British Film Institute.

——(1997) *Moving Images*, Manchester: Manchester University Press.

Castro-Thomasset, E. (2000) *L'apostasie de la television*, Paris: L'Harmattan.

Crain, M. M (1992) 'Pilgrims, Yuppies and Media Men', in J. Boissevain (ed.) *Revitalising European Rituals*, London: Routledge.

——(1996) 'Contested Territories', in J. Boissevain (ed.) *Coping with Tourism*. Oxford: Berghahn.

Davin, S (1999) *Casualty: Reception Study of a Medical Drama*, London: Le Drac.

——(forthcoming, 2003) 'Healthy Viewing: the Reception of Three Medical Narratives', in C. Seale (ed.) *Media and Health*. Oxford: Blackwell.

——(forthcoming, 2004) *La médecine dans le salon: Urgences et ses spectateurs*, Paris: L'Harmattan.

——(2004a) 'Public Medicine: the Reception of a Medical Drama', in K. Watson and M. King (eds) *Health and Illness in the Media*. London: Macmillan.

Dayan, D and Katz, E (1992) *Media Events*, Cambridge, MA: Harvard University Press.

Dowmunt, T. (ed.) (1993) *Channels of Resistance*, London: British Film Institute.

Eco, U (1985) *La guerre du faux*, Paris: Éditions Grasset.

Elkamel, F (1995) 'The use of television series in health education', *Health Education Research* 10 (2): 225–232.

Falk, J. H and Dierking, L. D. (1998) *The Museum Experience*, Washington, DC: Whalesback Books.

Fennell, D. A. (1999) *Ecotourism*, London: Routledge.

Fiske, J (1995) *Media Matters*, London: Routledge.

Fjellman, S. M (1992) *Vinyl Leaves*, Boulder: Westview.

Goffman, E (1974) *Frame Analysis*, New York: Harper and Row.

Hannigan, F (1995) The Post-Modern City *Current Sociology* 43, 1: 155–200.

Harrington, C. L. and Bielby, D. D. (1995) *Soap Fans*, Philadelphia: Temple University Press.

Hill, A (1999) *Shocking Entertainment*, Luton: University of Luton Press.

Hoijer, B (1999) 'To be an audience', in P. Alasuutari (ed.) *The Media Audience*, London: Sage.

Jenkins, H. (1992) *Textual Poachers*, London: Routledge.

Just, M. R., Crigler, A. N., Alger, D. E., Cook, T. E., Kern, M. and West, D. M. (1996) *Crosstalk*, Chicago: Chicago University Press.

Karpf, A. (1988) *Doctoring the Media*, London: Routledge.

Kepplinger, H. M. and Habermeier, J. (1995) 'The Impact of Key Events on the Presentation of Reality', *European Journal of Communication* 10 (3): 371–390.

Kottak, C. P. (1990) *Prime-Time Society*, Belmont: Wadsworth.

MacCannell, D. (1973) 'Staged authenticity: arrangements of social space in tourist settings', *American Sociological Review* 79: 589–603.

——(1992) *Empty Meeting Grounds*, London: Routledge.

Macdonald, S. (1995) 'Consuming Science', *Media, Culture and Society* 17: 13–29.

Maddox, B. (1999) 'How trustworthy is television?', *British Journalism Review,* 10 (2): 34–38.

Meyrowitz, J (1986) *No sense of place*, Oxford: Oxford University Press.

Morley, D. (1980) *Nationwide.* London: British Film Institute.

Morse, M. (1983) 'Sport on television', in A. E. Kaplan (ed.) *Regarding Television*, New York: American Film Institute.

Philo, G. (1996) *Media and Mental Distress*, London: Routledge.

Poster, M. (1996) *The Second Media Age*, London: Polity Press.

Poynton, B. and Hartley, J. (1993) 'Male Gazing'. In M. E. Brown (ed.) *Television and Women's Culture*, London: Sage.

Press, A. L. (1991) *Women Watching Television*, Philadelphia: University of Pennsylvania Press.

Reilley, J. and Miller, D. (1997) 'Scaremonger or Scapegoat?' in P. Caplan (ed.) *Food, Health and Identity*, London: Routledge.

Rodaway, P. (1995) 'Exploring the suject in hyper-reality', in S. Pile and N. Thrift (eds) *Mapping the Subject*, London: Routledge.

Rojek, C. (1993) *Ways of Escape*, London: Routledge.

——(1997) 'Indexing, Dragging and the Social Construction of Tourist Sights', in C. Rojek and J. Urry (eds) *Touring Cultures*, London: Routledge.

Sherwin, R. (2000) *When Law Goes Pop*, Chicago: University of Chicago Press.

Smith, V. L. (1989) 'Introduction' In V. L. Smith (ed.) *Hosts and Guests.* Philadelphia: University of Pennsylvania Press.

Stewart, C., Lavelle, M. and Kowaltzke, A. (2001) *Media and Meaning*, London: British Film Institute.

Tufte, T. (2000) *Living with the Rubbish Queen*, Luton: University of Luton Press.

Tulloch, J. and Moran, A. (1986) *A Country Practice*, Sydney: Currency Press.

Turnock, R. (2000) *Interpreting Diana*, London: British Film Institute.

Urry, J (1995) *Consuming Places*, London: Routledge.

Wayne, M. (ed.) (1998) *Dissident Voices*, London: Pluto.

Winckler, M. (2002) *Les miroirs de la vie*, Paris: Éditions Le Passage.

Woolley, B. (1993) *Virtual Worlds*, Oxford: Blackwell.

Zolberg, V. L. (1996) 'Museums as contested sites of remembrance: the Enola Gay affair', in S. Macdonald and G. Fyfe (eds) *Theorising Museums*, Oxford: Blackwell.

14 Converging cultures; Converging gazes; contextualizing perspectives

Rhona Jackson

I was motivated to write this chapter to make sense of a 1998 study trip to Los Angeles, because despite the reasonable assumption that tourists' anticipation of an enjoyable experience will be fulfilled, on this occasion, the opposite pertained. The purpose of our visit was to contextualize teaching and learning, to consolidate our knowledge of the Hollywood film and television industry. But, the tensions of the converging cultures of tourism and the media effected responses that we found difficult to understand and express. We had not anticipated such culture shock, to feel so foreign in a land which is allegedly Britain's closest political ally, where English is the official first language, with people whose screen images we see every day. Our national, cultural, and self-identities were tested as we encountered another nation's understanding of self and Other. In turn, we found ourselves repeatedly relying on flawed information about a country we thought we knew. Our 'knowledge' was continually tested, proving ultimately to be based on illusion. Instead of strengthening understanding, this trip resulted in feelings of uncertainty, disappointment and disenchantment. As Crouch and Lübbren (2003: 6) discuss how:

> [m]aterially, tourism is visually represented as significantly physical, involving space, visiting particular 'concrete' places. Metaphorically, visual culture may construct ideas and desires of the experience of tourism, and of particular imagined places

so our visit to LA foregrounded the tension between the material and the metaphorical. This, therefore, is an attempt to theorize such a tourism experience.I have used the tourism principle here, firstly to 'revisit' our tourist destination, setting the scene by recounting our impressions; second, to relate the theoretical journey taken to explain the experience.

The Cultural Studies approaches of the mass culture debate, Marxism, and postmodernism helped locate the significance of Hollywood as culture industry, ourselves as consumers of cultural products, ideological diversity, and sense of identity. However, because of their contextual focus, none of these approaches accounted for personal experience. Yet, this was *my* experience, admittedly one shared with others, but one which obliged a centrality of the human response, and a conceptual approach which considered the individual as well as the socio-cultural. Thus, my ultimate theoretical venture was to explore theories of the gaze; to attempt to integrate the tourist gaze, drawn from Foucault's work on the power of surveillance, with the film gaze, developed from Freudian psychoanalysis, which simultaneously objectifies the media text whilst seeking a familiar image. I aimed to investigate the complex relationship that visual media texts have with place, space, and travel, and the tensions this creates for tourists imbued with such images when they encounter the reality on which those texts are based.

Los Angeles: the place, the space, the people

The differing cultural conventions of the Americans and the British were established on arrival at Los Angeles Airport, customs officers warning: 'You're English? Don't walk at night!' Driving from the airport, we saw the distant location of our hotel in Down Town, the district featuring LA's only skyscrapers. They seemed stark, sinister silhouettes against the twilight background of a vast, shabby, shantytown. This landscape differed from that of any British city. The space was vast, the sky big. The buildings that caught our attention *en route*, the cinemas, garages, and churches, appeared to signal the city's priorities. To make sense of the reality of Los Angeles, we immediately compared it to an early Hollywood film. It was like travelling towards the *Metropolis* of Fritz Lang's 1926 imagination.

A further virtual comparison illustrated the differences between the buzzing business district of daytime Down Town, and its night-time existence. Once bereft of those who work there, it was clear that this is a place where poor people live, the advice of the customs officers proving startlingly apt. The only people on the streets after 6 p.m. were the homeless and the British. In scenes comparable to post-apocalyptic films/TV programs about the end of civilization, where ragamuffin people live in sewers, emerging only at night, as in the late 1980s TV series, *Beauty and the Beast*, as dusk settled, the homeless appeared on street corners, and near pedestrian crossings.

Ninety-nine per cent of homeless people were African-American, one per cent white, contrasting with most other people we met, who were Hispanic. Yet, nowhere was their Spanishness publicly acknowledged, there being no apparent civic provision, no Spanish street signs, no public information in Spanish. To survive in Los Angeles, the city with the Spanish name, one must function in English. Just as we were naively unaware of the huge problem beggars are in the USA, or, at least, in Down Town, LA, we hadn't known this. Such material conditions are not part of its standard media (metaphorical) representation.

Our dealings with people entrusted with transporting us to places/sales people, confirmed differences in attitudes. The tour of the stars' homes, visits to The Chinese Theatre, the Warner Brothers' lot, and seats at two live TV recordings were arranged from the UK, and confirmed by 'phone on arrival. Yet, the relentless verbal sparring with drivers/sales people made us singularly weary. We were constant prey to those determined to take advantage of our foreignness, feeling repeatedly bullied to hand over more money than agreed. Only 'playing them at their own game' brought us respect. To the mingled amusement and horror of our students, all conversations with minibus drivers became loud, public, aggressive, and oh, so un-British! The only people with whom we had a genuinely pleasant relationship were the waiters in a local, all-American diner.

As we became increasingly aware of and disappointed with how impoverished was our knowledge of LA and its people, so it became clear that their knowledge of the UK was equally partial and sparse. For example, we were repeatedly asked how Britain was faring after 'Diana's death', and 'how are the boys doing?' It was six months since Princess Diana had died, but it was difficult to convey that the British nation was not still mourning to the extent perceived in the media on the day of her funeral. We were also bemused by the assumption that we would know how her sons were coping. Equally mystifying, and somewhat embarrassing in the telling, were our responses. We answered as if we *were* representative of a British nation still grieving, as if we *were* personally close to the Royal Family. It was as if it was our duty to know these things, not to let the (British) side down or disappoint (American) expectations. We were asked, too, about the Protestant/Catholic problems in England. Our explanation that, although some terrorist attacks had taken place on the British mainland, the preponderance of 'the troubles' was in Northern Ireland, was met with bewilderment, and baffled queries of: '*Northern* Ireland?' Finally, as soon as people knew we were British, they went out of their way to tell us how much they liked The Beatles.

Thus, as our knowledge of LA, in contemporary USA, was fractional, biased and invariably downright wrong, so the Americans' awareness of the UK paralleled our ignorance. Our understandings of Hollywood, the place and society, proved inadequate, being based on screen images promoted by Hollywood, the culture industry. Correspondingly, the perception of Britain/British society prevalent in LA depended on British people/events with global impact; concerning royalty, terrorism, and popular culture. Rather than interest in our experience of the late 1990s British North Midlands, they related to a dead, although globally iconic, royal princess, to geographical misunderstandings about British religious differences, and OK, admittedly to the most famous rock group ever, but one which had disbanded nearly thirty years previously.

The sights and sites of Los Angeles

Certain tourist places also thwarted expectation. The stars' homes were fascinating, solely because we found them architecturally disappointing, due, presumably to British familiarity with the (often decaying) grandeur of ancient architecture associated with 'old money' in the UK, and so, consequently, our subconscious expectations of what grandeur is were not met. The only two points of 'specialness' were that these homes belonged to stars, and the separateness of their Beverly Hills' location.

Visiting the Chinese Theatre brought the realization that whenever this location is filmed, for instance, in the past for the Oscars ceremony, we see the stars, the red carpet, and the adoring crowd. Yet, immediately adjacent was wasteland, fronted by old railings sporting tacky advertisements, a bleak contrast to the TV-transmitted glitzy glamour. Media selection and editing conventions ensure that this image never reaches our screens. The interior décor of The Chinese Theatre was also far less interesting than that of the decaying 'Universal Cathedral' cinema in Down Town. Unlike The Chinese Theatre, this cinema's upkeep had long been abandoned, its location presumably determining its rejection for consideration for glamorous broadcasting projects.

But, where our expectations of the material reality of Los Angeles were frustrated, many of those of the mediated/metaphorical reality were fulfilled by the trip to Warner Brothers. After the post-production sound effects studio, the *Friends* set, the vast wardrobe warehouse, and the set for *ER*, walking to the woods from *The Waltons,* rubbing shoulders with Tom Selleck *en route*, we turned a corner to find our students playing cowboys and Indians on the set of *High Noon* and *Silverado*. Where the material sites had disappointed, they recognized and wanted to be part of the mediated/metaphorical, virtual reality of the film sets.

Finally, our seats in the audiences of two 'live' TV chat shows, *Politically Incorrect* and *The Tonight Show with Jay Leno*, confirmed our cultural differences from most others in the studio audience. It was a wise tactic to seat us far from the stage, presumably to avoid televising a non-American response to the programme, the audience in US TV being considered part of the 'atmosphere' of the programme/product. Our visible embarrassment at the loud, enthusiastic whoops and hollers from our American counterparts confirmed our stereotypically grim British reserve. Cultural references also meant little to us. Whilst recognizing the entertainers' words, we failed to grasp their meaning. We felt very foreign.

We were disconcerted that the only place we liked on sight was Santa Monica, its European 'feel' in terms of architecture and town planning obliging us to acknowledge our 'Little European' mentality.

We made sense of most places by comparing them to screen images, being surprised at our own naivety, having failed to appreciate how extensively our imaginations had been fuelled by the 'metaphoric construction of ideas and desires' (Crouch and Lübbren, 2003: 6). So, to rationalize such an unsettling experience, I turned to theory, beginning with the mass culture debate, whose perspective focuses on the cultural industries that manufacture film/television products.

Valuing culture

According to Real, the mass culture debate suggests that '[c]ulture. [is] life made up of symbolic forms in structured contexts' (1996: 35), the traditional values of artistic taste and creativity which inform those symbolic forms having been neglected in favor of product and profit. Similarly, Strinati defines popular culture as 'mass culture produced by mass production industrial techniques, marketed for profit to a mass public of consumers' (1995: 10). The focus is on product over art; teams of workers as opposed to an individual creator; industrial techniques not artistic inspiration and talent; marketed to rather than appreciated by; profit over aesthetic appreciation; vast numbers of consumers, whose *raison d'être* is to purchase and possess, rather than an elite, interested in taste and knowledge. The mass culture debate would explain our experience, by arguing that, except for a few outstanding examples, Hollywood films/TV programmes meet standards which accord with the formulaic, bland, repetitive nature of the mass culture debate's understanding of the reason for their production.

Moreover, Real maintains that '[i]deologically, Hollywood is one of the major instruments of the capitalist-consumer status quo' (1995: 152). Our experience of the way those working in the culture industry behaved, whose *only* interest was in taking our money, confirms the argument that they act in ways which the ordinariness, familiarity, and vulgarity of mass culture encourages. Being treated solely as brainless consumers was horrible. And the understanding of the audience as 'atmosphere' in the production of the TV programmes, rather than as appreciative, social beings, underlined the notion of people being useful only for their contribution to/place in/promotion of product. This was not merely about the contrast between people working in the tourist industry, and others pursuing entertainment and pleasure as tourists. Whilst it did illustrate conflicts between work and leisure, product and art, it also highlighted the way behavior expresses attitude, and the tensions which arise when opposites collide.

However, there are problems with using the mass culture debate. First, it denies that any value can be attributed to popular culture. Yet, I was in LA exactly because I did value certain popular cultural products, namely films/television programmes made there. Second, the British perspective on the mass culture debate is that Americanization is culturally damaging. So, to apply it to the American culture industry in the USA seems wholly inappropriate (and awfully discourteous). Third, this approach assumes the mass, passive audience to lack that discernment implied by the traditional use of the term 'culture'. As a member of that audience, strangely, I don't go along with that!

However, considering the significance of the ideological differences between the Americans as host and ourselves as tourists, the Marxist perspective promised clarification of ideological influence on socio-cultural encounters.

LA makes Marxism user-friendly

And Althusserian Marxism did shed some light on our feelings of exclusion. Althusser argued that ideology has a material existence in lived experience, notably in terms of Ideological State Apparatuses (ISAs), such as the mass media, and popular culture; that ideology expresses an imaginary relationship to our real conditions of existence; and that ideology 'interpellates', or calls to individuals as subjects of ideology (1971: 227–228). The material existence of ideology was manifest by the way the mass media dominates Los Angeles, promoting the dominant (metaphoric) ideas circulating in popular culture, influencing/working with the other ISAs of education, the Church, and the family to uphold the ideology of the American Dream, via Hollywood, 'the dream factory'.

The differences between the material reality of LA, and the virtual/metaphoric reality of those screen images that powered our expectations pointed up the imaginary relationship we had to the real conditions of existence of material people living in contemporary USA. Our understandings were structured by what we saw on screen, with no material evidence to anchor them. In fact, material evidence challenged the metaphoric of our imagination. Meeting people whose aim was to 'make a buck', rather than be glamorous and glossy media image portrayals, was just disappointing.

Grossberg *et al*'s argument that '[p]eople are interpellated by other people's language and behavior, as well as by the media texts that address them' (1998: 228), implies that interpellation assumes understanding and acceptance of cultural convention, whereas we felt excluded, and alien. We were not interpellated by American ideology as subjects of that ideology, but as subjects of British ideology. We brought UK cultural/ideological values based on lived experience and applied them to LA, making value judgements that found the landscape, architecture, and behavior of local inhabitants wanting. So Althusserian Marxism clarified that feelings of difference and alienation had ideological roots, and explained the ideological influences of media texts. However, it ignores any pleasure gained from film/television texts. Yet, despite seeking an explanation for my negative tourist experience, my reason for going to LA was exactly that pleasure I had gained from those cultural products which had come out of it in the first place.

Consequently, because of its focus on style over substance, a postmodern approach seemed apt in a city where the principal *raison d'être* is to (re)produce celluloid/technological images of reality, Hollywood appearing to confirm Mitroff and Bennis' argument that everything and everywhere 'is now a sub-branch of the entertainment division of the world' (1989: 53), the trivia of entertainment taking precedence over everything else, as substance is ever subordinated to style.

Too much style and not enough substance?

Particularly pertinent here were the notion of 'simulacra' (Baudrillard, 1983b), the focus on the consumer, and debates around identity. Baudrillard conceptualized the simulacra, where representations of material reality substitute for just that material reality, to the extent that so many copies have been made of copies that an original no longer exists, and 'the ability to distinguish reality from its images (has) disappear(ed)' (Grossberg *et al.*, op. cit.: 55). That many of our expectations of Hollywood were met by the Warner Brothers visit supports this. We related far more easily to the simulated reality of the mediated production/metaphoric reality of the USA than to the material reality which (once) informed it.

Similar to the mass culture debate, postmodernism also focuses on the consumer. We visited Hollywood as consumers of cultural products, our roles as consumers being emphasized both as members of two TV programmes' studio audiences, and as participants in the construction of audience-as-commodity (ibid. 215). However, because we felt excluded by both programmes' cultural address, we failed to engage with them on any level; as consumers of product, audience for entertainment, or participants in commodity production. This served to return our critical distance, resulting in a failure to acknowledge the 'product' as desirable. The strangeness of the entertainers' conversations thwarted any intended inclusion by our roles as part of the 'atmosphere', our alienation and awkwardness undermining the success of the audience-as-commodity.

Postmodernist debates concerning identity were useful because they raise questions about the definitions and boundaries of self/social/ cultural/national identity, and how the mass media influences/contributes to understandings of same (Strinati 1995; Real 1996; Sarup 1996), which did indeed impinge on our experience, underpinning our discomfort when excluded from the host cultural universe. So, a postmodern approach offered insight into debates around material/metaphoric reality, consumerism, and identity. However, once again, it could not account for those *feelings* which are key to tourist experience.

The mass culture debate's focus on product, and consumerism helped, but its denial of value, disdain for everything American, and negative conceptualization of the audience meant it was theoretically inadequate to explain my tourism experience fully. Althusserian Marxism grounded the experience in ideological difference, but could not account for the fundamental reason for our visit and its failure, the dimension of pleasure/unpleasure. And postmodernism, whilst introducing the material/mediated reality relationship, and bringing notions of identity into the equation, was not equipped to fathom the emotional response central to the tourist experience. The theoretical parameters of these Cultural Studies approaches, therefore, usefully permit analysis of socio-cultural experience, but constrain consideration of those whose socio-cultural experience it is. Theories of the gaze, however, do incorporate the notion of self.

Reviewing the gazes

Exploring theories of the gaze shifts the focus from the place/product to the person and what they do with that place/product. Plus, both tourism and film studies have developed concepts of the gaze to hypothesize how/why tourists/spectators view spectacle, and what they derive from looking. Notions of pleasure, and emotional response can, therefore, be explored.

Both tourists and film spectators/television viewers gaze. Tourists gaze at material reality, which in contemporary society is fundamentally influenced by mediated, virtual, metaphoric reality. Film spectators/TV viewers gaze at mediated, virtual, metaphoric reality, based on contemporary understandings of material reality.

The tourist gaze

Urry (1990) conceptualized the tourist gaze as 'constructed in relationship to its opposite', to explain how tourists understand the visited place and people as the Other, by consuming and owning what they purposefully gaze at from a position of power, which has been ideologically granted by the destination. The tourist gaze appropriates Foucault's (1976) ideas of surveillance and power in terms of the medic/patient relationship. The medic visits, is permitted to gaze from a position of power and ease at the patient, whose ideological position is of social powerlessness, dis-ease, the gazed at. Medic and patient understand the medic to represent the ideological norm; the patient is the Other, who deviates from that norm. The tourist gaze thus incorporates notions of scrutiny and surveillance, terms that imply deliberate, detailed examination over time. Although not discussing the medical gaze in depth, Urry maintains that the tourist gaze conforms to Foucault's model. The tourist imposes ideological understandings on the visited place and people, thus, setting the norm.

Implicit here is that the tourist gaze is a conscious operation of ideological power. As the medic understands that knowledge gained from study/experience grants power over the patient, so the tourist's knowledge of their own culture, and previous experiences of other cultures, give them power over the destination, to (re)view that destination as the Other. Curtis and Pajaczkowska explain that travel 'combines the pleasure of displacement with the enjoyable role of ethnographer/consumer and the positions of heightened authority which accompany the power to totalize and appropriate' (1996: 201). The pleasure of being away from home, of inhabiting different space, experiencing alternative place, is enhanced by the knowledge that the tourist holds ideological sway over their destination, the place and space of the Other.

I suggest that the tourist gaze is, therefore, underpinned by the identity of the tourist. To retain ideological ease the tourist relates to the visited place and people just as Sarup contends the dispossessed and lost search for identity, 'in terms of preoccup[ation with] ideas of home, displacement, memory and loss' (1996: 93). The tourist constantly compares what is gazed at with what is familiar. As Sarup explains the foreigner (ibid: 99–100), so the tourist 'enter[s] the culture of others but ... this move is undertaken only to return to oneself and one's home, to judge or laugh at one's peculiarities and limitations' (ibid. 100), and, equally, I contend, to judge or laugh at the peculiarities and limitations of the tourist destination.

Because that's what we do as tourists. We visit another place, we compare it to home and everyday surroundings, and sometimes we find it wanting. For instance, Donald (1997), corresponding with Crouch and Lübbren's later argument concerning material and metaphoric tourism, suggests that the tourist experience includes physical visiting, and, crucially, the imagination, where '[t]o imagine is to make present to ... [one's] ... mind's eye what is absent' (ibid. 183). And Urry argues that '[t]he tourist gaze is directed to features of landscape and townscape which separate them off from everyday experience' (op.cit. 6), whilst it seeks the familiar, which has been:

> constructed and sustained through a variety of non-tourist practices, such as film, TV, literature magazines, records, and videos, which construct and reinforce that gaze [which has been] constructed through signs, and the tourist is a collector of signs.
>
> (ibid. 3)

Taking Donald's use of the term 'imagine' to mean the incorporation of anticipation and curiosity, framed by expectation, this is an exercise in surmising, based on accrued knowledge. As Urry points out, just as tourists take with them their 'home ideology', so they bring also their mediated experience of the tourist destination, amassed in terms of signs. To undertake such comparison, they use many cultural tools and resources, including language and the media.

We did all these things in LA. We compared the imagined tourist experience with the actual, and with conventions of home. Our USA tourist experience was framed by our UK lived experience, and by 'knowledge' of US life, accumulated from a succession of film/television images about the USA, broadcast in the UK. Curtis and Pajaczkowska (1996: 206) suggest that this comparison of differences between cultural activities, social behaviour, perceived identities, and the framing of these experiences by mediated screen images, repeatedly test, and ultimately reinforce the ideological power which cloaks the tourist, 'impl[ying] a circular confirmation of self-identity'.

Thus, the concept of the tourist gaze introduces the self into the explanation of our tourist experience of LA. However, it barely acknowledges the significance of the relationship between the tourist and those who inhabit the tourist destination. Urry notes economic and social differences which may be evident (1990: 58), but does not explore the feelings these may engender on either part.

A central concern of film theory, however, is the relationship of the spectator to the mediated image. I suggest, therefore, that there is a way to bring together the tourist and film gaze, so that the converging cultures of tourism and the media form an integrated conceptual approach. This would incorporate an explanation of how the gaze operates for the tourist both in terms of place and space, and in images of place and space, as well as bringing the person into the equation, that self whose gaze is being explored.

Accepting that film spectators/TV viewers can be described similarly to Urry's definition of tourists, as 'collector[s] of signs', and that in contemporary society the tourist is likely also to be a film spectator/television viewer, it is reasonable to argue that the concept of the tourist gaze could be developed by appropriating certain elements of the film gaze. The tourist gaze seeks the familiar, which relates in part to the conceptualization of the film gaze. However, in contrast to the concept of the tourist gaze resting on critical distance, its premise being the power of surveillance, the film gaze is motivated by a desire to seek (the security of) intimacy, to pursue a closer relationship.

The film gaze

Film theorists, such as Christian Metz (1975) applied Freudian psychoanalysis to the way a spectator experiences a film, arguing that Freud's (1905, 1917) explanation of the child's relationship to 'the imaginary' and 'symbolic' stages of development, could clarify the relationship of the spectator to the images on the cinema screen. Lacan's reworking of Freud (Mitchell and Rose 1983) to maintain that the unconscious is structured like language, and that the use of language prompts the perception of one's own subjectivity, contributed to Metz' findings.

Proposing that looking at a cinema screen is analogous to looking into a mirror, Metz suggested that watching a film replays the primary instances of perception and recognition; it repeats the first understanding and realization of the difference between self and image. The spectator's gaze at the screen is also a displaced fascination with the ideal self of the mirror-image. This in turn represents the security of the pre-Oedipal phase before self-awareness is achieved, understood as a repressed (regressive) desire to return to 'the imaginary' stage of development. But the relationship between spectator and cinema is substitute for, not replica of, the child/mirror relationship. What is ever-present in the latter, the *self*-image, is always absent from the former. Unlike the child, the cinema spectator is equipped with knowledge of his own subjectivity, Metz explaining that he understands both the object of his perception, and the process of his perception. 'I know I am perceiving something imaginary ... I know that it is I who am perceiving it' (1975: 51). He does not need the validation of self that is the self-image (sic).

The film spectator is simultaneously involved in the cinema screen image whilst distanced from it, because perception and recognition make up his own subjectivity; because of his existence in 'the symbolic', conscious state of being (sic). This paradox inflicts problems of identity on to the spectator. Looking at the screen becomes a search for the always absent self-image, whilst invoking a desire to 'lose oneself' in Other(s), in those images present on the screen.

Acknowledging these arguments, film theorists deemed it reasonable to assume that film is predicated on the desire to look/awareness of looking; and to explain the pleasures of looking at film by those pleasures Freud originally associated with looking. Narcissism, or the fascination with self-image, becomes for the spectator the fascination with looking at an image of one with whom they wish to identify. Voyeurism, the overwhelming curiosity of the child to see his parents' genitals/his parents having sexual intercourse, becomes the desire to look at others, when the person who looks is unseen and the persons looked at are (or behave as though they are) unaware of being seen. Exhibitionism, the enjoyment of being looked at, is invoked by identification with the image as an image-to-be-looked-at. Fetishism displaces the threat of castration symbolized by woman, by representing her simultaneously as the Other, and as phallic substitute.

These are the appropriations film theory made from psychoanalysis. I now venture a further appropriation, to transpose them on to the concept of the tourist gaze, to clarify the search for the familiar, the nature of the power of the gaze and of those who gaze, and the understanding of the self-who-does-the-gazing. The self-awareness of the film spectator, as foregrounded by Lacan and Metz, is recognized as key to the gazer who is the tourist.

Converging gazes

My argument here is that tourism, too, is predicated on the desire to look, and to possess (by appropriation) what is looked at. In film theory, this desire is motivated by the psychoanalytical determination of male hetero-sexual gratification, whereas the tourist desires to look in order to feed the power associated with socio-culturally understood superior knowledge of place and space, and the ideological permission of the visited.

In terms of narcissism, the tourist, like the film spectator, seeks a familiar image, someone like themselves, or one with whom they wish to identify. For us, this was thwarted because those with whom we associated in terms of how they looked, reversed our expectations of how the Other should behave. By assuming a mantle of power we had not granted, our expectation of setting the norm was denied, the expected power relationship reversed as we were treated as the Other. As their roles contradicted their images with which we expected to identify, this would explain some of our uneasiness.

If we then remove from the concept of voyeurism the key Freudian issue of sex, although inevitably reducing its power in a 'pure' psychoanalytic sense, what is left is the contemporary, 'popular' meaning of the term: objec-tification by looking. The power and ease rests with the tourist who gazes, the discomfort/dis-ease with those gazed at. This was highly applicable to our LA tourist experience. We were most comfortable when looking at the 'virtual reality' of the celluloid/technological products/metaphor of the screen images of Hollywood stars and locations with whom/which we were most familiar. We were irritated when material people resisted the discomfort associated with the gazed at, and failed to conform to our knowledge of mediated/virtual people.

Exhibitionism is less useful as a concept, unless applied to tourists' awareness of being tourists, of their engagement with the 'strangeness' of the tourism destination, of wanting to be photographed, for instance, with 'locals'. (However, I suspect that this is stretching this term to make it fit, rather than applying it effectively.) Fetishism can be understood as displacing the power associated with the objectified visited, by representing them simultaneously as the Other, and as a potential source of pleasure for visitors. The above explanations in terms of narcissism and voyeurism support this.

Where being a tourist and being a film spectator/television viewer could, therefore, be argued to be converging cultural activities, the gazes theorized to explain such activities can be integrated to provide deeper insight. Both theories of the gaze make it possible to explain at least some of those personal feelings aroused in the tourist by individual encounters with the indigenous population of the host destination. My argument is that such feelings should be deemed as significant an aspect of the tourist experience as the appreciation of place and space.

Central to this is that imagination and expectation determine how we understand reality, and our identities in that reality. It allows reality to resist concrete definition, and our understanding of it to assume a slippery shifting from lived experience to mediated image to imagined comparisons of the two. It becomes less and less clear whether/how the material informs the metaphorical/virtual, or vice versa, emphasizing the complex relationship between the physical and its representation, and the tourist experience of it/them. Our visit to LA demonstrated that when the two collide, disappointment may ensue.

As Cronenberg's film, *Videodrome* (1983), states, 'television is reality and reality is less than television... .' And this testing of reality foregrounds uncertainties relating to notions of identity, be it self, socio-cultural, or national. Furthermore, where the tourist gaze implies that the gazer keeps a critical distance, to survey all that they behold, entailed in the film gaze is that the person who gazes is driven by/surrenders to the need to gaze, in obsessive pursuit of that 'self-image'. Combining the two approaches strengthens the theory of the gaze, as it weaves the subconscious drive of the film spectator into the ideological but knowing aloofness implicit in the tourist gaze. It establishes the self as theoretically significant, permitting the search for self, in terms of all sorts of related debates, to be understood to be as fundamental to the tourist gaze as it is for the film gaze.

Contextualizing perspectives

The mass culture debate, Althusserian Marxism, and postmodernism cannot explain the individual, as they are approaches developed to explore society, the collective. The tourist gaze, drawing on Foucault's understanding of how power works in society, establishes a theoretical perspective that reveals one particular way in which such power can operate, providing a bridge from approaches to society/culture to theories of the individual, by acknowledging the significance of what the tourist does to gain tourist experience. The film gaze developed from Freudian psychoanalysis, and is specifically a theory of the individual self. Permitting the two gazes to work together, and contextualizing them in terms of the Cultural Studies approaches means that the individual being, the socio-cultural being, and society in general can be considered from a combined conceptual, albeit complex and eclectic, perspective. It results in an approach that authorizes the place of 'humanness' in the tourist experience to be central to socio-cultural theory. It acknowledges and offers a way of explaining the complex feelings derived from visits to other places, explorations of different space, and encounters with people, all entailed in the tourist experience. Thus, both positive and negative tourism experiences can be investigated and explained.

There are five provisos associated with the above argument. First, implicit here is that the film gaze is equally applicable to television. Much writing in this area contests this (cf. e.g. Ellis: 1982). However, I contend that, as long as we remain aware that this theory developed to explain film spectatorship, then, with care, it can indeed be applied to television viewing.

My second point may seem a strange omission from a woman writer, but I plead the excuse of constraints of time and space. It is important explicitly to acknowledge the 'problem' that the theory of the film gaze is premised on a norm of male heterosexuality. However, the consideration of how gender and sexuality influence the tourist gaze as well as taking account of the many feminist film theorists on the 'male gaze' (cf. e.g. Mulvey 1975; Doane 1982; Kuhn 1982) cannot be done justice in a mere section of a chapter.

Third, it is worth reiterating that all theories are partial. The mass culture debate exposes the influences of cultural industries; Althusserian Marxism stresses ideological subjectivity; postmodernism foregrounds the nature of a society bound by media images, consumerism, and the search for identity. None account for the self. The theories of the tourist gaze and the film gaze insert what the tourist does, including dimensions of self and personal perspective. Yet, they provide little explanatory satisfaction in terms of contextual constraints. Only by combining them in an integrated approach can a comprehensive conceptualization of the tourist experience be established.

However, my fourth point is that considerable care must be taken to apply such an approach. Its 'openness' provides a great advantage for the analyst, permitting a comprehensive understanding of the tourist experience as so many influences from a number of perspectives can be taken into account.

Conversely, it has the potential of creating a monster, an approach too awkward and unwieldy to effect. Its successful operation obliges a precise, concise, and clear framework of expectation, followed by rigorous methodological application, and account.

Finally, I do not maintain that the two theories of the gaze, contextualized by Cultural Studies approaches, combine with automatic, unified ease. On the contrary, although both conceptualizations are premised on looking, and the other approaches provide respective socio-cultural perspectives, there are considerable theoretical tensions, due in no small part to philosophical origins, and initial aims. I suggest, rather, that they be understood to work in parallel, not necessarily entirely harmoniously, to offer a deeper insight into what it is to be a tourist in a society driven by mass media images.

References

Althusser, L. (1969) *For Marx*, Harmondsworth: Penguin.

——(1971) *Lenin and Philosophy and Other Essays*, London: New Left Books.

Anderson, B. (1991) *Imagined Communities* (2nd edn), London: Verso.

Baudrillard, J. (1983a) *In the Shadow of the Silent Majorities*, New York: Semiotext(e).

——(1983b) *Simulations*, New York: Semiotext(e).

——(1983c) 'Consumer Society', in M. Poster (ed.) *Selected Writings*, Stanford, CA: Stanford University Press.

Billingham, P. (2000) *Sensing the City through Television*, London: Intellect Books.

Couzens Hoy, D. (ed) (1992) *Foucault: A Critical Reader*, Bristol: Blackwell.

Crouch, D. and Lübbren, N. (2003) *Visual Culture and Tourism*, Oxford: Berg.

Curtis, B. and Pajaczkowska, C. (1996) 'Getting there: travel, time, and narrative'. In M. Sarup (ed), *Identity, culture and the postmodern world*, Edinburgh: Edinburgh University Press.

Doane, M. (1982) 'Film and the Masquerade: Theorizing the Female Spectator', *Screen*, 23 (3–4), 74–78.

Donald, J. (1997) 'This, Here, Now: Imagining the Modern City', in S.Westwood, and J. Williams (eds) *Imagining Cities: Scripts, Signs, Memory*, London: Routledge.

Ellis, J. (1982) *Visible Fictions*, London: Routledge and Kegan Paul.

Foucault, M. (1976) *The Birth of the Clinic*, London: Tavistock.

Freud, S. (1974) [1917] *Introductory Lectures on Psychoanalysis*, Harmondsworth: Penguin.

—— (1977) [1905] *On Sexuality*, Harmondsworth: Penguin.

Grossberg, L., Wartella, E. and Whitney, D. C. (1998) *MediaMaking*, London: Sage.

Kuhn, A. (1982) *Women's Pictures: Feminism and Cinema*, London: Routledge and Kegan Paul.

McHoul, A. and Grace, W. (1993) *A Foucault Primer*, London: UCL Press.

Metz, C. (1975) 'The Imaginary Signifier', *Screen*, 16 (2): 14–76.

Mitchell, J. and Rose, J. (1983) *Feminine Sexuality: Jacques Lacan et l'Ecole Freudienne*, London: Macmillan.

Mitroff, I. I. and Bennis, W. (1989) *The Unreality Industry*, New York: Oxford University Press.

Mulvey, L. (1975) 'Visual Pleasure and Narrative Cinema', *Screen*, 22 (2): 29–42.
Real, M. (1996) *Exploring Media Culture*, London: Sage.
Sarup, M. (1996) *Identity, culture and the postmodern world*, Edinburgh: Edinburgh University Press.
Strinati, D. (1995) *Introduction to Theories of Popular Culture*, London: Routledge.
Urry, J. (1990) *The Tourist Gaze*, London: Sage.
——(1995) *Consuming Places*, London: Routledge.

Films/television programmes

Beauty and the Beast, 1987–90, dir. D. Attias and F. Beascoechea, USA.
ER, 1994 to date, dir. M. Crichton, USA.
Friends, 1994–2004, dir. M. Crane and M. Kauffman, USA.
High Noon, 1952, dir. F. Zinneman, USA.
Metropolis, 1926, dir. Fritz Lang, USA.
Politically Incorrect with Bill Maher, 1997–2002, Brad Grey Television, USA.
Silverado, 1985, dir. L. Kasdan, USA.
The Tonight Show with Jay Leno, 1992, dir. E Brown, USA.
Videodrome, 1983, dir. David Cronenberg, USA.
The Waltons, 1972–81, dir. W. Altzman and G. Arner, USA.

15 Producing America

Redefining post-tourism in the global media age

Neil Campbell

'Stay in and go somewhere different' (Sky Television advertisement 2002)
'[I]n the world of supermodernity people are always, and never, at home'
(Augé 1992:109)
'A virtual America, therefore, would be a mythic America turned inside out'
(Giles 2002:14).

This chapter examines how as tourists in the global media age our perceptions of the USA are influenced by both the amount of cultural information repeated and circulated and what we do with it. There has been much debate about the nature and scope of Americanisation as a one-way process of cultural imperialism infiltrating the lives, both conscious and subconscious, of individuals and nations, with tourism as one source of this perceived 'takeover' (see McKay 1997; Tomlinson 1991; Campbell, Davies and McKay forthcoming). Whether from tourism to the USA itself, or simply by engaging with American-style tourist experiences and practices, some would argue that there is a powerful redefinition taking place, one often paralleled with the impact of fast food or theme park culture in the way that it regulates and manipulates behaviour (see Ritzer and Liska in Rojek and Urry 1997). Thus so-called 'McDisneyization' is a specific example of cultural imperialism showing how tourists can be considered part of the Americanisation process imposing its rules, values and ideologies upon a relatively passive consumer. This monologic process is interrogated in this chapter, allowing for the possibilities that these 'consumers' are simultaneously 'producers' too, working *with* the mediated 'America' they encounter in all manner of subtle ways to construct something different and more multiple than is often considered under the assumptions of Americanisation. Therefore, out of the interplay of debates over tourism, or 'post-tourism', and Americanisation one might reassess the nature of travelling in ways that contest many of these assumptions and provide a different, expanded and more productive version of cultural identity in the global media age.

Jennifer Price's *Flight Maps* opens with the question 'What does nature mean to me?' and then searches for it in the most obvious and unexpected places, scattered, as it is, almost everywhere (1999: xv). Gathering up these discursive formations of 'nature' she assembles a record of her 'travels', as she calls them, through all manner of experience; 'travels at once unsettling and reassuring', mapping the different ways nature is constructed and produced by us all (ibid. xxii). Price's use of the word 'travel' suggests her discovery of 'nature' is a form of tourism experiencing multiple, mediated spaces, both material and immaterial, past and present, real and imagined from which she assembles and constructs her meanings for nature, which in turn, contribute to the formation of her identity and sense of place. Similarly, one might ask, 'What does America mean to me?' Where do I go to find it? How and where do I 'discover' and 'produce' America? Can I find it without leaving home?

The conclusion, like Price's, is that the itineraries of everyday life bring us into contact, wherever we are, with America as real and imagined space, seeing it from the outside, as mediated, simulated, mythic *and* as actual, lived and tangible in life's everyday and multi-layered spaces. These complex geographies produce America's 'scriptural economy' (de Certeau 1988: 132): from the fast food restaurant's slogan of the 'United Tastes of America', to the shopping mall's Disney, Warner Brothers' or Timberland stores, to the theme park's time-space compression of American experiences from the Alamo to space travel, to the array of texts and images available on-line. Thus we can 'know' America as media tourists despite living far away and having limited, actual contact with the nation, rather like Barthes' sense of Japan in *Empire of Signs*, a 'fictive nation' constructed by a 'number of features ... deliberately form[ing] a system ... which I shall call: Japan' (Barthes 1982: 3). Out of the 'faraway' signs that under globalisation are ever-closer at hand, Europe assembles a 'system' called 'America' in a process not unlike what Augé terms the 'anthropology of the near' (Augé 1992: 7) where a city centre stroll, a theme park visit, a shopping trip, a meal in a restaurant, a glance at a billboard, magazine, guide book, or merely surfing the Internet brings a mediated America into our lives less as an exotic destination 'out there', than an immediate, negotiated, and even contested, tourist experience.

The postmodern media surrounds us, as in Barthes' Japan, with fragments, narratives and representations that as tourists we incorporate or reject as a sense of place is formed. Such overlapping touristic sensibilities enter into the way we structure our lives making it less possible to think of tourism as a 'discrete activity contained tidily at specific locations and occurring during set aside periods' (Franklin and Crang 2001: 7). Indeed, those moments that in the past might have been considered as discrete tourist experiences marked by a search for authenticity away from home and work have become diffused into much of our day-to-day lives marking the 'shifting boundary of holiday and everyday' (ibid.). Through this 'tourism of everyday life' people are 'routinely excited by the flows of global cultural materials all around them in a range of locations and settings' without having to physically travel (ibid. 8). Aided by immense advances in the electronic media we 'casually take in these flows' in ways once only experienced through physical travel, learning new repertoires, acquiring different expectations and translating them into the day-to-day world as a complex mélange of actual and virtual experiences (see Virilio 1998: 67; Shields 2003).

These ideas redefine post-tourism in terms of virtual/actual travel without the necessity of a faraway place, distinct from work, positing tourism as the elaborate consumption and use of a series of images in what I will term the virtual construction of America. The once singular activity of the tourist, seeking an authentic experience away from the workaday world, has been replaced by a gamut of experiences, knowledges, anticipations, activities, and performances that constitute a postmodern tourism both self-aware and hybrid, mixing 'texts' into an elaborate weave of leisure practices distinct from a 'pre-packaged' holiday. Post-tourism in this manner contests traditional notions of tourist experience offering more than physical travel including, as it does, desire, imagining and mediation in a much more complex and encompassing mobility close to what James Clifford calls 'dwelling and traveling' (Clifford 1997: 30). In 'travelling' to America through its 'cultural flows' and the mediated 'virtual' Americas of television, advertising, consumerism, the service industries, film, literature and the Internet, it is possible to engage actively in the construction of a real and imagined nation or 'transnation' whilst still 'dwelling' in a familiar and rooted existence. As John Rajchman reminds us, 'To virtualize nature [*or* America] is thus not to double it but, on the contrary, to multiply it, compli-cate it, release other forms and paths in it' (Rajchman 1998: 119).

The passive Americanisation of experience associated with cultural impe-rialism's idea that 'The Media Is American' (Tunstall 1977), is challenged by this new tourism's active, performative practice through which the post-tourist encounters America as objects, places, simulations, fashions, pleasures, tastes, and sights/sites creating a tourist text as a:

> galaxy of signifiers, not a structure of signifieds; it has no beginning; it is reversible; we gain access to it by several entrances, none of which can be authoritatively declared to be the main one; the codes it mobilizes extend as far as the eye can reach, they are indeterminable.
>
> (Barthes 1975: 5–6).

These 'mobilized codes' defining 'America' impose no simple cultural–imperialist model upon the audience since they have been actively producing (and contesting) meanings through their consumption. The new post-tourist neither receives a complete, monologic package called 'America', nor engages in the impossible quest for an authentic singular 'true' America, but rather constructs and inhabits a contingent, playful, fragmentary space, both real and imagined, confirming and challenging rooted, mythic discourses of America associated with such concepts as freedom, space, diversity, the frontier, progress and individualism. This self-reflexive, participatory tourism has, in a sense, been tutored through our very consumption of the media and our sophisticated 'use', pleasure and critique of its many elements and 'codes'.

For example, when Ien Ang famously studied the television programme 'Dallas' she concluded that rather than accepting a pre-packaged set of values and expected assumptions, its audiences had various responses, entering into complex dialogues with the show proving 'it is wrong ... to pretend that the ideology of mass culture exercises dictatorial powers [since] alternative discourses do exist which offer points of identification' beyond those expectations (in During 1993: 416). With the media, as with tourism, people 'practise' it variously, mixing, contesting, combining and pluralising their use of its outputs creating all forms of pleasure, meaning and identity. As David Crouch reminds us, 'people use different agendas from those supplied by commodification and work their own sense into these leisure spaces' since space, like identity, is not totally controlled by others, but is always 'practised', adapted and hybridised in all kinds of productive ways (Crouch 1999: 257–8). Thus a show like *Dallas*, which many felt was Americanising the world, could actually function in juxtaposition, alongside other mediated experiences, as an integral part of a tourist's critical 'work' constructing a 'Dallas–Texas–America nexus', acting precisely as an 'invention and interruption of meaningful wholes in works of cultural import-export' (Clifford 1988: 147).

The American mythic 'package', like the Grand Canyon as we shall see later, supposedly sold in advance to tourists might, therefore, be undone by this more variable, dialogic relationship to its representations and mediated 'script', for as Clifford argues, it is 'a hooking-up and unhooking, remembering and forgetting, gathering and excluding of cultural elements – processes crucial to the maintenance of an 'identity' – [that] must be seen as both materially constrained and inventive' (Clifford in Gilroy *et al*, 2000: 97). Out of this 'uncomfortable site' of constraint and invention the new post-tourist might 'begin' to chart a different identity through its ambivalent experiences of America's mediascape while simultaneously resisting and challenging the wholesale, passive acceptance of Americanised cultural values (ibid.).

Redefining the post-tourist

The term 'post-tourist' was coined by Maxine Feifer responding to a street at Mont St Michel in Normandy full of 'creperies, Coca Cola stands, and tourist boutiques selling gimcrack souvenirs' where the tourist had 'come all this way to see something venerable, beautiful, and above all different [to find only] an atmosphere of other tourists: the modern plight' (Feifer 1985: 2). The post-tourist has learned to live with and enjoy this 'modern plight' as part of the tourist repertoire within a highly mediated environment: 'Via the mass media, one knows a little bit about a lot of things' (ibid. 260). Standing by the Eiffel Tower she recalls and quotes Barthes' famous essay's idea that what one 'sees' is 'the most general human image-repertoire ... confronting the great itineraries of our dreams' triggered in advance by its media saturation and iconic presence into which one is inserted as tourist identity. Feifer's experience radiates outwards from Barthes with 'wry hyper-self-awareness', to include buying souvenirs, tourist chatter, and generally revelling in the 'touristy' 'simulated environment' rather than evading or denying it, recognising that the mediated experience is indeed integral to the nature of tourism (Feifer 1985: 267, 269). She adds that 'As the McLuhanesque global village of communications media gets bigger and more elaborate, the passive functions of tourism (i.e. *seeing*) can be performed right at home, with video, books, records, TV' (ibid. 269).

Looking beyond Urry's influential notion of the 'tourist gaze' (1990) as the determining element of tourism, she calculates the effects of the 'overabundance of [media] events' on the totality of the tourist experience (Augé, 1995:30). The 'playful' post-tourist *in situ* 'traverses a landscape' noting its 'geometric complexities ... jazzlike discordances [and] variety of aesthetic contexts' with the 'humorous eye for "kitsch" as well', enjoying 'the connective tissue *between* "attractions" as much as the vaunted attractions themselves' (Feifer 1985: 270 – my emphasis), but in addition calls upon a vast range of 'other' media-generated 'landscapes' of sensation, knowledge and imagination that, in the words of Arjun Appadurai, 'transform the field of mass mediation because they offer new resources and new disciplines for the construction of imagined selves and imagined worlds' (Appadurai 1996: 3). This is at the heart of reconstituted post-tourism's potential for contestation and transformation, blending the actual and the virtual as components of the twenty-first century leisure event into 'a constitutive feature of modern subjectivity' (ibid.).

Feifer's post-tourist is creative rather than passive in receipt of the defined and pre-packaged experience cast out from the 'nets of the media' (de Certeau 1988: 165), knowing 'that he is a tourist: not a time traveller when he goes somewhere historic; not an instant noble savage when he stays on a tropical beach; not an invisible observer when he visits a native compound' and is, therefore, able to 'embrace' and critique it as part of the process (Feifer 1985: 271). Feifer recognised that achieving 'authenticity' was an archaic tourist desire related to an impossible belief in the 'real' and the 'original' experience, now clearly altered (and enhanced) irredeemably by the omnipresence of the media in all its forms. Her examples help us realise there can be no authentic, true tourist experience since even the 'authentic' is a 'cultural construction' sustained by 'ideologically infused discourses and delusions' (Lewis, 2003: 6), and that ultimately the media reminds us that tourism is a series of simulations from which we build our own 'package'.

In the eighteen years since Feifer's book there has been a media proliferation perpetuating and accelerating opportunities for the types of post-tourist experiences she charts, especially in the growth of digital satellite technologies and the Internet, capable, as Augé writes, of conveying 'an instant, sometimes simultaneous vision of an event taking place on the other side of the planet' (Augé 1992: 31). These mediated spaces enter lives with rapidity, compressed in time, 'as a substitute [or *supplement*] for the universes which ethnology has traditionally made its own' (ibid. 32). Just as ethnology is redefined in the light of these changes, so is tourism, accommodating the overabundance of mediated events and knowledge to actively produce a virtual/actual experience. Operating within a version of Augé's 'non-place', both everywhere and nowhere, the post-tourist combines the imagined (a dream of America, media representations, screen cultures), the 'real' (actual travel, guides and themed experiences) and the virtual (myths, media, Internet) into a 'package', a collage-like America of over-lapping and disjunctive elements that together construct their tourist experience. The post-tourist's playfulness is creative, translating identities between place and non-place 'like palimpsests on which the scrambled game of identity and relations is ceaselessly rewritten' (ibid.:79). The post-tourist is no simple detached 'reader' of the pre-scripted tourist text, but an active 'writer–reader–practitioner' using the skills of everyday life such as multi-tasking, rapid interchanges and shifting between media forms and communication flows to assemble, produce and consume virtual 'Americas' that together form a multi-layered collage cutting up and rearranging any given 'scriptural economy'. To borrow from Appadurai, this tourist-media intersection provides 'resources for self-imagining as an everyday social project' choosing and moving across and between various texts inventing America as 'a mass-mediated imaginary that frequently transcends national space' (Appadurai 1996: 4, 6).

The potential of such mediated tourism is partially defined by Michel de Certeau:

> Without leaving the place where he has no choice but to live and which lays down its law for him, he establishes within it a degree of plurality and creativity. By *an art of being in between*, he draws unexpected results from his situation.
>
> (de Certeau 1988: 30 – my emphasis).

When de Certeau lists the media saturation of the modern world and asks 'What do they do with it?' he directs us back to our central question too ... what do we do with all these American images? Following de Certeau, the new post-tourist is no Americanised, passive being, but mixes and matches from the *mediascape* forming their own virtual America, part mythic cliché, part surprising and hybrid invention, part narrative debris (ibid.107). Gathering up these mediated fragments of place and identity, a new 'collage' is created that denies the oft-made assumption that the tourist-as-consumer is a 'sheep progressively immobilized and "handled" as a result of the growing mobility of the media as they conquer space. The consumers settle down, the media keep on the move' (ibid.165). In fact, the post-tourist 'keeps moving' but differently, crossing boundaries, shifting liminally between experiences without necessarily having to travel in any conventional manner (see Shields, 2003: 13). The 'Americanised' tourist, 'grazing on the ration of simulacra the system distributes' (ibid.166) perpetuated in texts from Eco and Baudrillard to Ritzer and encapsulated in the concept of McDisneyization (see Rojek and Urry 1997), is a reductionist vision that defines the consumer as a 'receptacle ... similar to what it receives ... passive, "informed", processed, marked, and [with] no historical role', when in reality they can be more inventive, *'travel[ling]* through ... texts' with 'detours, drifts ... produc[ing] by the travelling eye, imaginary or meditative flights' (de Certeau 1988: 167,170). In the twenty-first century one must extend de Certeau's 'reading' to global media texts that amplify this 'travelling' through which one constructs 'another world', since in the mediascapes that structure tourism 'readers are travellers; they move across lands belonging to someone else, like nomads poaching their way across fields they did not write ... his place is not *here* or *there*, one or the other, but neither one nor the other, simultaneously inside and outside, dissolving both by mixing them together' (ibid. 174).

The 'nomadic' post-tourist dwells and travels, moving and 'poaching' between the arrays of experiences that constitute 'America' actively constructing a hybrid sense of place and identity in the process. Rojek develops these ideas by emphasising the media's role in the post-tourist experience, adopting terminology drawn from personal computing and digital technology. He writes of the 'index of representations' – visual, textual and symbolic – on which the post-tourist might draw to construct their own desired landscape (Rojek and Urry 1997: 53). The archive's 'files' are, however, more than visual involving the post-tourist in a creative process of 'dragging' (as on an active desktop) as they are moved, combined, selected, deleted, in acts of 'interpenetration of factual and fictional elements to support tourist orientations' (ibid.); an elaborate 'copy', 'cut' and 'paste' across the mediated spaces of the contemporary world. Rojek draws upon the American example I used earlier to make his point, showing how Dallas could be 'indexed' and 'dragged' from diverse narratives and fragments such as the Kennedy assassination and the Ewing family in the TV show *Dallas*. To similarly access 'America' today involves an even wider selection of 'files' to draw upon, as we shall see with the Grand Canyon, engaging the post-tourist in 'fantasy-work, reverie or mind-voyaging' (ibid. 63), indexing and dragging no longer as 'an unavoidable accessory to the sight', but as integral to the experience. This mediated 'collage tourism' 'replaces [even] the necessity physically to visit the site' (ibid.) and presents, at its most positive level, a hybrid engagement between cultures and texts paralleling Clifford's 'ethnographic surrealism' with its capacity to 'mock and remix', invent and interrupt dominant cultural assumptions (Clifford 1988: 117).

Urry recognised that 'People are tourists most of the time whether they are literally mobile or only experience simulated mobility through the incredible fluidity of multiple signs and electronic images', and has developed this idea further in the concept of 'mobilities ... at the heart of a reconstituted sociology' (Urry 1995: 148; 2000: 210). Accepting the multiple nature of tourism in the digital age, concepts such as 'imaginative mobilities' and 'virtual travel' form part of this expanded vision in which 'It becomes possible to sense the other, almost to dwell with the other, without physically moving either oneself or ... physical objects' (ibid.:66, 70). Deploying the ideas of Gilroy and Clifford, for whom culture is about mixing routes *and* roots, Urry stresses the 'complex relationships between belongingness and travelling, within and beyond the boundaries of national societies. [Where] People can indeed be said to dwell in various mobilities...' (ibid.:157). A redefined post-tourism is equally 'mobile' in this sense, moving in the circuits of mediation both dwelling *and* travelling amidst the 'intercultural import–export' of everyday life, sifting and selecting experiences, producing mobilities and reinventing identities (Clifford 1997: 23; see Cresswell 2001). Indeed, Sky television's slogan 'Stay in and go somewhere different' might inadvertently signify this new tourist sensibility derived from an inventive use of these media possibilities.

Virilio's post-tourist inertia

For some however, such fluid ideas of dwelling and travelling within the media appear less optimistic, leading to claims of overt Americanisation, passivity or, as in Paul Virilio's work, to a world being 'shrink-wrapped by global media' as it 'substitutes' actual experience with the virtual, consequently diminishing the human in the accelerated saturation of new global 'vision machines' that 'overexpose' and synthesise physicality (McQuire in Armitage 2000: 146). As he puts it, the global media is creating its 'final resting-place ... shrinking before our eyes to a blind cockpit for the dreams of a population of sleepwalkers' (Virilio 2000: 31). This is an age characterised, according to Virilio, by 'a monotheism of information' bombarding the senses and distracting the individual from previously accepted notions of travel experience (in Armitage 2000: 13). Substituting for these experiences are the multiple 'screens' of the global media networks projecting into the individual's home until 'everything is on the spot, everything is played out in the privileged instant of an act, the immeasurable instant that replaces extension and protracted periods of time' making that individual a 'tele-actor' detached from 'physical travel' taking on 'another body, an optical body' that will 'go forward without moving, see without eyes, touch with other hands than his own, to be over there without really being there, a stranger to himself, a deserter from his own body, an exile for evermore' (Virilio 2000: 17, 85).

The experience of tourism involving motion across the earth is altered fundamentally by this 'telluric contraction' since the media now replaces and reconstructs 'the very nature of our travels' making it 'travelling on the spot, with an inertia that is to the passing landscape what the 'freeze-frame' is to the film' (ibid.:18). Virilio's virtual or *cyber* travel transmits electronic mediascapes through 'vision machines' into the everyday spaces of the home creating 'the static *audiovisual vehicle*, a substitute for bodily movement and an extension of domestic inertia which will mark the definitive triumph of sedentariness' (ibid.). With no capacity for human choice, interaction, empowerment or pleasure, these new digital media epitomise 'a sort of Foucauldian imprisonment [in which] the world is reduced to nothing [and so] it is no longer necessary to go towards the world, to journey.... Everything is already there' (Virilio in Armitage 2000: 40). For Virilio the port of entry and departure of classical tourism has been translated into the 'teleport' in the heart of the inert home creating tourism without travel as a substitute for real life and meaningful interaction: '*Now everything arrives without any need to depart* [through] the general arrival of images and sounds in the static vehicle of the audiovisual. Polar inertia is setting in' (Virilio 2000: 20–1; italics in original).

Virilio relates this process of 'substitution' to the 'Americanised' theme park; a denudation of experience, a space of inaction, media detachment, and the epitome of cultural loss:

> The leisure park is on the point of becoming a stage for pure optical illusions, a generalization of the non-place of simulation with its fictitious journeys offering everyone electronic hallucination or intoxication – a 'loss of sight' replacing the nineteenth-century loss of physical activity.
>
> (Virilio 2000: 19)

Virilio's nightmare world of loss can be countered by defining the traveller-as-*agent*, following de Certeau, inventing and producing tourist 'texts' as a 'collage', working with and through the mediated landscapes either as virtual in themselves, or in relationship with the actual, physical acts of travel in a process of supplementation not of substitution. To use our example of apprehending America, Virilio might interpret this as a manifestation of negative global substitution as people absorb the second-hand representations, myths and narratives through screen technologies of sameness, converting different spaces and cultures into a hideously monologic 'world city', a 'geostrategic homogenisation of the globe' (McQuire in Armitage 2000: 147). Virilio's tendency is to essentialise an authentic, earlier sense of identity being eroded by the forces of hypermodernity in similar ways to the notion of the ideal 'classical' tourist or traveller, with a fixed agenda of motives and principles. However, as we have seen, one can view this differently, with the media contributing to a hybrid post-tourism that 'indexes' and 'drags', selects and rejects America, as part of a complex gathering-up process that constructs a rather more fluid, non-essentialised and mobile identity.

Themed environments and the Grand Canyon

To demonstrate this post-tourism I will examine two inter-linked examples of this projected 'America', through themed environments such as malls and restaurants and as iconic sites, like the Grand Canyon. Both can be visited and experienced as 'American' in multiple ways as we 'index' and 'drag' their representations and actuality into our own personal vision. Theming is pervasive, occupying spaces beyond the real and actual, existing through the media circulating images and experiences of America; in the UK as 1950s retro culture in the 'OK Diner' chain of Streamliner-styled aluminium restaurants with sporting memorabilia themes, or as the mythic Wild West at The American Adventure theme park with its simulated saloon, dancing girls and shoot-outs juxtaposed with the Alamo and Native American remnants, or in any trip to the shopping mall (Gottdiener 2001).

Many critics view these themed experiences as hollow spectacles showing all the attributes of a McDonaldization or McDisneyization of tourism (see Ritzer and Liska in Rojek and Urry 1997; Eco 1986; Baudrillard 1983 and Virilio 2000). Eco, for example, views Disneyland as 'a disguised supermarket, where you buy obsessively' and for Baudrillard it is 'there to conceal the fact that it is the "real" country, all of "real" America, which *is* Disneyland' (Eco 1986: 43–46; Baudrillard in Storey 1993: 164). Thus in visiting British shopping malls, like the Trafford Centre, Meadowhall, or Bluewater, we enter 'a stunning new world of experience' like a theme park, taking us to New Orleans or New York City through its themed restaurants and attractions, 'the choice is ours' (Trafford Centre brochure), learning to be post-tourists dipping in and out of a new psycho-geography of American fragments. As David Harvey puts it, summarising many of these criticisms of themed mall cultures, this is a 'time–space compression' wherein, just as 'all the divergent spaces of the world are assembled nightly as a collage of images upon the television screen' so theme parks allow us 'to experience the world's geography vicariously, as a simulacrum' whilst 'conceal[ing] almost perfectly any trace of origin, of the labour processes that produced them, or of the social relations implicated in their production' (Harvey 1990: 300). Alternatively, these experiences become part of a creative repertoire the individual draws upon, often reflexively and critically so that the 'collage' is not inevitably, as Harvey suggests, reductionist.

After all, we *know* it is not 'real' America, since we have not crossed the Atlantic, and yet we still sample the iconic, mythic 'themes' that holiday brochures and television programmes articulate. For example, having shopped in The Gap, eaten at McDonalds and walked through a simulated New Orleans, we could buy an American magazine or novel or DVD, or read a USA holiday brochure whose opening double-page has images of happy Hawaiians, the Statue of Liberty, a cowboy and the words 'Big, Bold and Larger than Life ... *Imagine* the soaring skyscrapers of Manhattan, the glittering lights of Las Vegas, glamorous Hollywood legends and the cowboy country of the West. Think of the great outdoors: awesome Niagara Falls, the majestic Grand Canyon ...' (*Funway* 2003–2004). The mediated, iconic 'themes' are here once again and we are required to 'imagine' them before (or as well as) actually experiencing them, drawing them out from our already-formed sense of America and reassembling them as a new collage – 'a vast melting-pot of creeds, colours and traditions', where, as with the shopping mall, 'the choice is yours' (ibid.).

As we assemble our particular 'America' of themes, icons, micro-narratives, instant histories and mediated images we *consume* and *produce* as post-tourists, as de Certeau argued, willingly indulging in the McDisneyization of leisure fused with the realities of everyday existence. Some would argue, like Ritzer and Liska, that such a process of predictable, efficient, calculable and controlled experience – the defining elements of a McDonaldized society – have thus entered tourism and the media, with the inevitable outcome that 'real', authentic travel (*and* identity) has disappeared and been substituted with 'virtual travel' and hollowed-out identity: 'some people will find that it is far more efficient to 'visit' Thailand [or America] in the comfort of their living rooms [or themed environments] than actually to journey there' (in Rojek and Urry 1997:101; Virilio 2000). These negative visions of endless, passive consumption cohere in the vision of a 'McWorld' of malls, multiplexes, theme parks, fast-food chains and television, forming a huge enterprise transforming and denuding humanity (see Barber 1995). A different approach asserts active choice, 'sampling' and 'collage tourism' within this process of consuming places and experiences whereby the McWorld's inauthenticity and commodification is acknowledged and 'used' pleasurably in the construction of a post-tourist America as a series of critical dialogues *with* America, rather than the passive recipient of a pre-formed monologue.

These complex dialogic relations to themed and mediated tourism can be translated into responses to an iconic American experience like the Grand Canyon – the so-called 'seventh wonder of the world' – where theming has spilled over into the organisation of nature as 'regulators ... designers and engineers have worked over natural wonders ... to heighten the theme of mother nature in an idealized sense' (Gottdiener 2001:3). But even before arrival, assuming one is going at all, the Grand Canyon 'travels' into the post-tourist consciousness via its intense and varied media presence. Recently, for example, a BBC viewers' poll invited people to vote for the best place to visit 'before you die' and first choice was the Grand Canyon followed by Las Vegas and New York in the top ten places (www.bbc.co.uk/50/destinations/america). Although clearly a popular 'real' tourist destination, America also exists as 'imagined' spaces recalled from a cultural image-archive rich in specific mythic traditions about America as 'free', 'open', adventurous and vibrant. In these tourist experiences the 'America' discovered is, as the brochure stated earlier, 'larger than life', somehow more real than reality, a spectacular theme-park capable of taking us out of ourselves. The BBC website carries comments that emphasise the Grand Canyon's vastness and its exoticism drawing upon its 'natural', and, therefore, authentic presence: 'The Paiute Indians call it Kaibab ... "Mountain Lying Down" ', whilst simultaneously endorsing it with celebrity tributes from Eamonn Holmes to Jilly Goolden, and the authority of the *Condé Nast Traveller*. In turn the BBC web-site links to many others about the Grand Canyon; in fact a simple Google search produced well over a million web sites that draw one into the ultimate post-tourism journey.

Without moving further than a computer screen we engage with both 'official' and private sites on river running, air trips, hiking trails, wild life, environmental groups, Teen Summer Camps, gift shops, photo-galleries, individual travelogues from all over the world, and can even sit and watch the 'live' webcam of the actual Grand Canyon as it shifts and changes throughout the day and night. In 'indexing' and 'dragging' through this wealth of material a mediated Grand Canyon assembles following the 'quick links', cutting in and out of the 'frequently asked questions', the web cam, 'facts/docs', maps and news releases until one, ironically, fulfils, without moving, the National Park Service's slogan – 'Experience Your America'.

Alongside the computer screen, other media representations contribute to this post-tourist archive, such as Lawrence Kasdan's film *Grand Canyon* (1992) in which the 'real' place functions both as an image of social distance and a reassuring metaphor of 'otherness' for the lives of people in Los Angeles trapped in a world of crime, consumption and crises: 'there's a gulf in this country, an ever-widening abyss between the people who have stuff and the people who don't have shit, like this big hole has opened up, as big as the Grand Canyon'. At the end of the film the diverse characters take a trip to the actual place, reflecting as they gaze upon the canyon, 'I think ... it's not all bad', suggesting the mythic landscape's transcendental qualities to erase difference and to assert an ahistorical, harmonising and healing presence that one also discovers in Carl Sandburg's Prologue to the 1955 photography exhibition 'The Family of Man', describing it as 'A camera testament, a drama of *the grand canyon of humanity*, an epic woven of fun, mystery and holiness – here is the Family of Man!' (Steichen, 2000:5 – my emphases). Alternatively one might experience a sensational IMAX film or view the 'factual' video 'Reader's Digest Grand Canyon' (1988) inviting us to 'experience [its] awesome beauty and magnificent drama ... the splendour ... and magic ... in the comfort of [our] ... own home ... in living colour ... with specially scored stereo music ... and enlightening narration ... [to] Achieve *the feeling of actually being there*!' (my emphases). Wherever one discovers the Grand Canyon, as part of a virtual America, we are confronted by its real and imagined presence from which we negotiate and produce mobile meanings that interfere with and destabilise old, mythic, idealised visions as we confront, analyse and rearrange these elements into new patterns and juxtapositions.

Post-tourism, as it 'indexes' and 'drags', is less about defining separate cultures such as a monolithic, dominant America, or in asserting fixed and stable identities, but is engaged in producing 'conjunctures ... complex mediations of old and new, of local and global' accommodating the 'shifting mix of political relations' where nations and identities meet, intersect and hybridise (Clifford in Gilroy *et al* 2000: 98, 102). In this case, the Grand Canyon, as post-tourist event, samples these elements in their simulated virtuality *before* and *alongside* any actual physical experience of the geography of place; indeed, this is at the very heart of, what we might term, 'conjunctural' tourism, the liminal space in which varied and dialogic impressions and experiences meet, coalesce and interpenetrate.

'Every story is a travel story' (de Certeau 1988: 115)

This chapter has examined relations between Americanisation, media and tourism, but resisted seeing cultural imperialism as uni-directional because, as Raymond Williams has written, any society 'is *an active debate and amendment* under the pressures of experience, contact, and discovery, writing themselves into the land ... made and remade in every individual mind' (in Highmore, 2002:93 – my emphasis). America is indeed 'made' and contested within such 'active debate and amendment' without controlling identities or smothering 'indigenous' cultural forms and ideas. The actual, material landscape reminds us of this fact as we move *between* Americanised shopping malls *and* traditional marketplaces, visit theme-parks *and* museums, as the mixtures and 'impurities' around us testify to the cultural, spatial conjunctures that reflect the formations of identity constituted within this climate of addition, juxtaposition, supplementation and hybridity. Unlike Virilio's pessimistic vision of travel turned by the media into 'polar inertia', Homi Bhabha's sense of cultural identity as complex, hybrid and contested is closer to this chapter's arguments:

> What is at issue is the performative nature of differential identities: the regulation and negotiation of those spaces that are continually, *contingently*, 'opening out', remaking the boundaries, exposing the limits of any claim to a singular or autonomous sign of difference ... where difference is neither One nor the Other but *something else besides, in-between* ... a form of the 'future' where the past is not originary, where the present is not simply transitory.
>
> (Bhabha 1994: 219)

This 'in-between' is, as we have seen, the condition of the post-tourist of, say, the Grand Canyon, and is a productive rather than a reductive experience, actively 'performative, deformative', translating 'America' through mediated and actual tourism, re-thinking assumptions and expectations in a process of combination not substitution and of 'opening out' rather than of 'closing down' (ibid.241, 227). Rejecting a monolithic vision of America, the post-tourist plunges into a productive experience of time-space with no 'story-line unfolding sequentially' or 'ever-accumulating history marching straight forward', but with events 'happening against the grain of time ... *continually traversing the story-line laterally*' (Soja 1989: 23 – my emphasis).

The 'traversing' post-tourist enters a phase of what Edward Said calls 'overlapping territories, intertwined histories' where relations between Europe and America are no longer seen as one-way or essentialised, but rather as collaborative, negotiated 'contrapuntal ensembles' stressing 'a more urgent sense of the interdependence between things' (Said 1993: 1, 60, 72). These transnational 'connections' emphasise dialogue and contestation paralleling the reconsiderations of tourism and travel in a global 'post-cultural imperialist' media age where

> No one today is purely one thing. Labels like Indian, or woman, or Muslim, or American are no more than starting-points.... Imperialism consolidated the mixture of cultures and identities on a global scale. But its worst and most paradoxical gift was to allow people to believe that they were only, mainly, exclusively, white, or black, or Western, or Oriental. Yet just as human beings make their own history, *they also make their own cultures and ethnic identities* ... Survival in fact is about the connections between things ...
>
> (ibid. 407–8 – my emphases).

To borrow an idea from Rajchman in his discussion of the relations between the actual and virtual, I would conclude by suggesting that these concepts of tourism, media and identity are like constructing of a new type of house 'that holds together the most, and most complicated, 'different possible worlds' in the same container, allowing them to exist together along a constructed plane with no need of a preestablished harmony' (Rajchman 1998: 117). As we have seen, post-tourism's construction of a virtualised America rejects simple mythology or pre-packaged national narratives that engulf 'real space' or subsume identity and Americanise them both, and instead offers an alternative emphasis upon a

> virtual construction ... that frees forms, figures, and activities from a prior determination or grounding ... allowing them to function or operate in other unanticipated ways; the virtuality of a space is what gives such freedom in form or movement. Thus virtual construction

departs from organizations that try to set out all possibilities in advance. It constructs a space whose rules can themselves be altered through what happens in it.

(ibid. 119)

In this dwells a powerful sense of potential rather than loss, an assertion of new identities over a yearning for lost essences and of being 'always more than this actual world, and not limited by its already present forms' (Colebrook 2002: 96). As Giles argues, this process states a vital critical position, for the 'redescription of American culture as a virtual construction would seek to position itself on [geographical and intellectual] boundaries and, by *looking both ways*, to render the mythological circumference of the nation translucent' (Giles 2002: 14 – my emphasis). Thus the post-tourist has much in common with other diasporic or nomadic groups in the twenty-first century whose inventive mobilities move them *in-between* worlds unsettling the assumptions of one culture from the perspective of another, providing new ways to imagine identity and nation, ways that might, ultimately challenge us all to see beyond monolithic, mythic conceptions of closed cultures to something more mutable, itinerant and contested.

Works cited

Appadurai, Arjun (1996) *Modernity At Large: Cultural Dimensions of Globalization*, Minneapolis: University of Minnesota Press.

Armitage, John (2000) *Paul Virilio: From Modernism to Hypermodernism and Beyond*, London: Sage.

Augé, Marc (1992) *Non-Places: An Introduction to an Anthropology of Supermodernity*, London: Verso.

Barber, B. R. (1995) *Jihad Vs. McWorld: How Globalism and Tribalism are Reshaping the World*, New York: Ballantine.

Barthes, Roland (1975) *S/Z*, London: Jonathan Cape.

——(1976) *Mythologies*, London: Paladin.

——(1982) *Empire of Signs*, New York: Hill and Wang.

Baudrillard, Jean (1983) *Simulations*, New York: Semiotext(e).

Bhabha, Homi (1994) *The Location of Culture*, London: Routledge.

Campbell, Neil (2003) *Landscapes of Americanisation*, HEFCE/FDTL Project, University of Derby booklet.

Campbell, N., Davies, J. and McKay, G. (eds) (2005) *Issues in Americanisation and Culture*, Edinburgh: Edinburgh University Press (forthcoming).

Certeau, Michel de (1988) *The Practice of Everyday Life*, Berkeley: University of California Press.

Clifford, James (1988) *The Predicament of Culture*, Cambridge: Harvard University Press.

——(1997) *Routes: Travel and Translation in the Late Twentieth Century*, Cambridge: Harvard University Press.

——(2000) 'Taking Identity Politics Seriosuly: 'The Contradictory Stony Ground', in Gilroy, Grossberg and McRobbie (as below), pp.94–112.

Colebrook, Clare (2002) *Gilles Deleuze*, London: Routledge.

Cresswell, Tim (2001) 'The Production of Mobilities', *New Formations*, 43, Spring, 11–25.

Crouch, David (ed.) (1999) *Leisure / Tourism Geographies: Practices and Geographical Knowledge*, London: Routledge.

During, Simon (ed.) (1993) *Cultural Studies: A Reader*, London: Routledge.

Eco, Umberto (1986) *Travels in Hyperreality*, London: Picador.

Feifer, Maxine (1985) *Going Places: The Ways of the Tourist from Imperial Rome to the Present Day*, London: Macmillan.

Franklin, Adrian and Crang, Mike (2001) 'The trouble with tourism and travel theory?', *Tourist Studies*, 1 (1) 5–22.

Giles, Paul (2002) *Virtual Americas: Transnational Fictions and the Transatlantic Imaginary*, Durham and London: Duke University Press.

Gilroy, Paul (1994) *The Black Atlantic: Modernity and Double Consciousness*, London: Verso.

Gilroy, Paul, Grossberg, Lawrence and McRobbie, Angela (eds) (2000) *Without Guarantees: In Honour of Stuart Hall*, London: Verso.

Gottdiener, Matt (2001) *The Theming of America: American Dreams, Media Fantasies, and Themed Environments*, Boulder, CO: Westview Press.

Harvey, David (1990) *The Condition of Postmodernity*, Oxford: Blackwell

Highmore, Ben (ed.) (2002) *The Everyday Life Reader*, London: Routledge.

Lewis, Nathaniel (2003) *Unsettling the Literary West: Authenticity and Authorship*, Lincoln, NE: University of Nebraska Press.

McKay, George (ed.) (1997) *Yankee Go Home (And Take Me With You)*, Sheffield: Sheffield Academic Press.

McQuire, Scott (2000) 'Blinded by the (Speed of) Light', in Armitage, John (as above), pp.143–160.

Price, Jennifer (1999) *Flight Maps: Adventures With Nature in Modern America*, New York: Basic Books.

Rajchman, John (1998) *Constructions*, Cambridge: The MIT Press.

Ritzer, George and Liska, Allan (1997) '"McDisneyization" and "Post-Tourism": Complementary Perspectives on Contemporay Tourism', in Rojek and Urry (see below), pp. 96–109.

Rojek, Chris (1993) *Ways of Escape: Modern Transformations of Leisure and Travel*, London: Macmillan.

Rojek, Chris (1997) 'Indexing, Dragging and the Social Construction of Tourist Sights', in Chris Rojek and John Urry (see below), pp. 52–73.

Rojek, Chris and Urry, John (eds.) (1997) *Touring Cultures*, London: Routledge.

Said, Edward (1993) *Culture and Imperialism*, London: Vintage.

Shields, R. (2003) *The Virtual*, London: Routledge.

Soja, Edward (1989) *Postmodern Geographies: The Reassertion of Space in Critical Social Theory*, London: Verso.

Steichen, Edward (2000) *The Family of Man*, New York: Museum of Modern Art.

Storey, John (1993) *An Introductory Guide to Cultural Theory and Popular Culture*, London: Harvester Wheatsheaf.

Tomlinson, J. (1991) *Cultural Imperialism*, London: Continuum

Tunstall, Jeremy (1977) The Media Is American: Anglo-American Media in the World, London: Constable.

Urry, John (1990) *The Tourist Gaze*, London: Routledge

——(1995) *Consuming Places*, London: Routledge.
——(2000) *Sociology Beyond Societies*, London: Routledge
Virilio, Paul (1998, 1997) *Open Sky*, London: Verso.
——(2000, 1990) *Polar Inertia*, London: Sage.
Williams, Raymond (2002) 'Culture is Ordinary', in Ben Highmore (as above), pp.
    91–100.

16 Journeying in the Third World

From Third Cinema to Tourist Cinema?

Felix Thompson

Tourism and political cinema

How should the tourist analogy be handled in dealing with representations of Third World culture in film and television? Should the sense that the moving image caters for the tourist gaze be seen as fatally contaminating for any kind of progressive or political engagement? Or can the presence of a tourist dimension suggest ways in which metropolitan audiences are implicated in what appears to be a distant set of problems or a distant way of life? A film or television programme, once associated with tourism, is readily construed as manifesting diametrically opposite values to those of radical politics in the Third World. Yet it will be argued here that there are significant overlaps between Third Cinema as a political cinema of the Third World and relations of virtual travel and the tourist gaze. Initially, the practice of Third Cinema, which developed as a militant cinema in the context of Third World struggles against colonialism and neo-colonialism during the 1960s, appeared to be a long way from any kind of tourist gaze indeed. However, Third World cinema has in general moved away from the openly militant anti-imperialism of the 1960s.[1] This piece will address ways in which the tourist gaze and the political discourses of Third World film have come to intersect. The intersection will be considered from the point of view of the metropolitan spectator, who might be expected to be the most susceptible to the depoliticising effects of the tourist gaze.

A key concept here is the notion of virtual travel generated by the moving image media. The effect of virtual travel is shared in cinema both by openly touristic encounters with Third World cultures and more political versions of Third World film. The notion of virtual travel has been used to extend the complex negotiations of identity, space and place associated with the effects of travelling cultures as explored by James Clifford (1992). The media become an important means for understanding the relations of movement between cultures. But equally, for Clifford, travelling cultures are relevant within any specific culture to its internal relationships and sense of precise locality. Travelling cultures as experienced through the media pose questions about our own identity as much as they raise question about our relationships with other cultures.

Yet reactions to the implications of such virtual travelling cultures are divided. Martin Roberts, for instance, discusses the documentary film *Baraka* (Ron Fricke, USA, 1992), which includes 'reverential treatment' of 'unspoiled' environments, aboriginal societies and religions alongside sequences illustrating manifestations of social misery and war across the contemporary world (Roberts 1998). This documentary without narration cuts from one location to another with the panoptic claim to represent the state of the world between hope and despair. From Buddhist ceremonies through stunning landscapes to shots of slums or the blazing oil fields of the First Gulf War, an equally important rationale for the film is the quality of the photography which transports the viewer effortlessly across space and time. Against the scenes of environmental destruction and evidence of social breakdown, an image of pre-industrial collective harmony is suggested by Buddhist monks and aboriginal dancers. Roberts argues that the film is an example of the 'commoditisation' of the ethnographic, given the scope of a 'coffee table globalism' (1998: 66). The Western spectator is set at a safe distance from both the pristine localities and societies which capitalism is destroying and from the effects of such destruction.

The risks of assuming such a distanced vantage point are considered by Yosefa Loshitzky. In discussing television as a virtual travelling culture she poses stark alternatives. Television may, through the promotion of cross-cultural interactions, act 'as a potentially liberating force which equalises global power relationships' (1996: 329). On the other hand:

> through its 'ambivalent gaze' on the Other, it may perpetuate the Third World's status as undeveloped, and primitive space whose alluring power is seductive only for powerful voluntary western travellers in quest of exoticism and leisure.
>
> (ibid.)

A key group of films relevant to this debate were produced by what has been called the Fifth Generation of Chinese Cinema, for about a decade from the middle of the 1980s.[2] These films are concerned with political themes in the aftermath of the demise of Maoism, yet many of them present boldly coloured, dramatic representations of Chinese peasant life, open to appropriation by the tourist gaze. This is the implication of an argument produced by Rey Chow concerning *Ju Dou* (Japan/China, 1990), co-directed by Zhang Yimou with Yang Fenliang. The significance of Zhang Yimou's name in particular is that he came to this film with an outstanding international reputation for his cinematic exploration of traditional settings. The response to international interest in exotic images of China is further underlined by the fact that the film was co-produced from Japan. International co-producers and audiences sought an image of a pre-industrial China which is almost as far from the contemporary Chinese urban life of the film-makers as it is from viewers in Japan and the West.

The narrative concerns the young wife of the cruel elderly owner of a dye works who revolts against his brutality by beginning an affair with his adoptive nephew. In staging this story of traditional oppression the film makes much of colours appropriate to the setting in the dye works. Rey Chow suggests that what Western audiences will look for in the film is something other than realism in representing the lives of Chinese peasants:

> This is the cultural labour of the 'third world' in the 1990s, in which the 'third world' can no longer simply manufacture mechanical body parts to be assembled and sold in the 'first world'. What the 'third world' has been enlisted to do also is the manufacture of a reflection, an alterity that gives (back) to the 'first world' a sense of 'its' freedom and democracy while it generously allows the 'third world' film to be shown against the authoritarian policies of 'third world' governments. But an 'alterity' produced this way is a code and an abstraction whose fascination lies precisely in the fact that it is artificial and superficial; as Baudrillard says, 'it is the artifact that is the object of desire'.
>
> (1995: 60)

Chow's argument conjoins two separable perspectives here. First the Western spectator is given the clearly contrived artefact, an 'object of desire' offering the visual excess of colour. The way in which metropolitan viewers' visual involvement is being solicited here resembles one version of the tourist gaze described by John Urry, that of the tourist seeking an idealised photographic representation which is far more important than the direct experience of reality (Urry 1990: 86).

The second perspective is that of the viewer's judgement from the secure vantage point of western freedom and democracy upon the politics of a Third World country. For Chow, these two perspectives are perfectly aligned in *Ju Dou*. Yet significantly in this film, like the example of *Baraka*, the idealised representations of a remote way of life are set at a 'safe distance'. The vantage point from which the construction of this world is encountered is never, itself, brought into question. A central argument here is that this is not always the case and that, in particular, the dynamic effects of virtual travel through the moving image may lead to a significant separation and friction between touristic and political perspectives. It is also important that Chow's argument requires that there is a political case which makes sense. The viewer is to be cast in the classic position of the western Orientalist who can successfully make judgements upon Third World politics. What will be argued here, though, is that, through effects of mutually critical relationships between the political and tourist discourses, the sense of security in the tourist gaze and the Orientalist claim to political knowledge may be brought into question.[3] To pursue this argument it is necessary first to examine the developing role of political discourses in Third World Cinema since the Third Cinema of the 1960s.

Third Cinema and travel

The practice of Third Cinema was theorised in the 1960s as a militant cinema, principally emanating from the Third World, which adopted guerrilla tactics to challenge the oppressive world order and, particularly, neo-colonialism. Fernando Solanas and Octavio Getino, the initial Argentinian theorists of Third Cinema rejected the first cinema of Hollywood as an enclosed or 'hermetic' experience which, even if it showed concern for social issues, never asked the audience to engage in the world beyond the cinema. (Solanas and Getino, 1983) Second Cinema was European art cinema, including the radical efforts of film-makers such as Jean-Luc Godard. Such cinema was too confined to the conventional circuits of distribution and exhibition to have the kind of political involvement required by the urgent issues of the Third World.

Third Cinema is perhaps best illustrated in this militant period during the world conflicts of the 1960s by Solanas and Getino's own three-part film, *The Hour of the Furnaces* (Argentina, 1968). The first part of the film takes the form of a critique of neo-colonialism in Latin America, using satire and documentary footage to emphasise the intense level of the social and political crisis, making connections between the case of Latin America and major world conflicts such as the Vietnam war (Stam 1987). One particular theme is the cultural and economic subservience of the middle class in Argentina to Europe and North America. This is underlined at the end of the first part of the film by a stunning montage of the kind of foreign advertising which has penetrated everyday life in Latin America, succeeded by an extended icon-like shot of the dead Che Guevara. The production process of the film was completely outside the norms of the mainstream entertainment industry. It was assembled and re-edited in the light of comments from militant groups such as trade unionists. In the circumstances of political terror at the time it was generally shown in semi-clandestine situations such as militant meetings.

How then did the practice of Third Cinema develop from the high points of anti-colonial and neo-colonial struggle and the confrontations over Vietnam of the late 1960s? Despite the impressive list of directly militant films, there have been two major problems for the persistence of a militant practice of Third Cinema. The first is the disappearance from the world stage of the kind of focus provided by the confrontations of decolonisation and the Vietnam War. Secondly, there is an internal problem in Solanas and Getino's conception of Third Cinema. This problem arises with a film-making practice which rejects the cultural domination of Europe and the United States, a rejection which I have suggested was central to *The Hour of the Furnaces*. The issue here is one of finding a space from which to reject the cultural domination of colonisation. In *The Wretched of the Earth* Frantz Fanon argued that it is necessary in the struggle against colonialism to turn back to the indigenous culture of the colonised. This is not so much to suppose a pure state of culture which preceded the contamination of the invader as to find an alternative view of the world, based upon indigenous culture which could provide a motivation for struggle against colonialism and neo-colonialism.[4]

During the militant period of film-making in the late 1960s examples of this kind of return to indigenous culture developed and it is through this tactic that the political discourses of militant Third Cinema films come to intersect with issues of virtual travel and the tourist gaze. The return to indigenous culture is exemplified by the 1969 Bolivian film *Blood of the Condor* (Jorge Sanjines, 1969). Despite coming from the Spanish speaking elite, the collective who made the film, Grupo Ukamau[5], led by the director Jorge Sanjines, set out to make films from the point of view of the indigenous peoples of Bolivia in their struggles against military and economic oppression. *Blood of the Condor* concerns the clash between Andean peasants and the United States 'Peace Corps', which is accused in the film of sterilising Indian women without their consent. The film is shot predominantly with a hand-held style strongly associated with documentary in the 1960s, lending an urgency to the political issues it addresses. The Quechua peasants' suspicion of the role of the Peace Corps leads to an attack on a clinic which they run, followed by punitive measures of the neo-colonialist military government against the villagers in which their leaders are rounded up and shot by the police.

The main time frame of the film begins with the journey of the wounded Ignacio and his wife Paulina to find medical help in the city. They are shown descending long winding roads from their village in the Andes, with Ignacio carried on a stretcher. They then take a lorry to the capital, La Paz. There, his brother Sixto, who has attempted to establish himself in the big city, takes up the desperate search for medical help. Flashbacks explain the circumstances which have brought about the police attack on the village. The village is shown to be following traditional ways – Quechua language, clothes, festivals, religious ceremonies – which are troubled by the intervention of outsiders in the form of the police and the presence in the locality of 'Peace Corps' members and their clinic. The unexplained infertility which has affected the local population preoccupies the meetings of village elders and soothsayers. Their deliberations to determine the cause of the malaise lead directly to the attack on the clinic.

The organisation of the narrative around the journey, first from the village in the hinterland and then in Sixto's search for medical aid across La Paz, is notable because such a framework has frequently been used as a means of representing relationships between significant sectors of society in many subsequent Third World films. These kinds of journeys, it will be argued, play a particular role in bringing into question the vantage point of the metropolitan spectator. The relations implied by such journeys may be inflected in certain significant ways. First, they can be used to construct a sense of cultural difference and interchange between hinterland and metropolis. Second, the journeys provide some kind of reflection upon tourist relationships and the tourist gaze, even if, as in the case of *Blood of the Condor*, the activity of tourism is totally rejected through the politics of the film. Third, such journeys are central to an understanding of varying political stances taken in the films. This becomes important as the apparently incompatible perspectives of politics and tourism have become increasingly intertwined in the development of Third World film.

The concern with cultural difference is exemplified in *Blood of the Condor* by the representation of Sixto's alienation in the city. He is introduced wearing contemporary metropolitan clothes and working in a factory. Playing a game of football after work he is cursed for being an Indian despite his emphatic denials. This self-alienation is rapidly challenged by the arrival of his wounded brother. In his brother's cause, he encounters the gulf between the plight of his family and the casual indifference both of city life and the middle-class doctors. At one point Sixto follows a woman in a tourist market, gazing at her handbag which might contain enough money to save his brother. The setting of the tourist market, with its mixture of the kind of traditional items which we have already seen in the village alongside knick-knacks such as cuckoo clocks from across the world, emphasises the effects of deracination of peasant culture which parallels Sixto's existence in the city. While waiting outside the doctor's house in a wealthy suburb we gaze with Sixto, in a state of extreme anxiety about his brother's health, at the bizarre and irrelevant leisure activities of the elite, shown playing tennis and swimming. The comfortable position for the tourist gazing upon traditional Quechua life is therefore undermined. Indeed, it is as if the gaze has been turned back on the very positions from which a tourist gaze might be mounted. Everyday activities familiar to the metropolitan viewer become a means of registering a sense of incommensurability with Sixto's peasant background. After his brother's death Sixto is shown back in the mountains in traditional clothes. The shift in his consciousness towards a militant political identification with the peasant cause is suggested by a final shot showing peasant arms raising rifles into the air.

Touki-Bouki and the phantasmagoria of Third World travel

The filming of *Blood of the Condor* in documentary style underlines, through techniques of verisimilitude, the political importance of the opposition between metropolitan centre and traditional hinterland as a means of making sense of a Third World country. Yet, even in scenes where cultural antagonism is being expressed, it is difficult to maintain an equal belief in both worlds. From Sixto's point of view the leisure activities of tennis playing and swimming seem like a strange fantasy, as do scenes of members of the 'Peace Corps' dancing to rock music in their clinic when shown from the point of view of the indigenous inhabitants. This problematic representation of the intermingling of the two worlds of tradition and metropolitan culture is explicitly addressed in *Touki-Bouki* (Diop Djibril Mambety, Senegal, 1973) rather than being contained within a uniform documentary style of shooting and reduced to stark political polarisation. In *Touki-Bouki* metropolitan culture and that of the hinterland are presented as a source of division within the contemporary psyche. The film is firmly located within the everyday present of the shanty town, almost with a sense of documentary actuality, but it works back to the role of the metropolis/hinterland couplet as it inhabits the minds of the protagonists. The handheld assertion of traditional authenticity in *Blood of the Condor*, which suggests that the culture of the hinterland is just there to be observed as an equivalent reality to that of modernity, is replaced by an exploration of the symbolic power of tradition in the face of the symbolic power of modernity.

Touki-Bouki concerns a young hippy-like couple, Anta and Mory, who have plans to escape their West African shanty town and follow their dream of travel to Paris. While the couple are located in everyday urban life, Mory, in particular, is shown to have memories of a childhood as a cattle herder in the countryside. This sense of Mory's origins is reinforced when the couple travel on an *Easy Rider*-style bike through the bush while, with obvious irony, the dream of a journey to the metropolitan centre is signalled by songs about Paris on the soundtrack. The trajectory of the relationship between the hinterland and a distant metropolis is thus replayed in the film, although the two poles of the projected journey are rendered through memory and fantasy. Before they finally try to board the ship for France the couple are shown imagining their return with all the trappings of the fortune they will be able to bring back. Yet Mory is haunted by memories of a traditional way of life represented by his aunt Oumi who holds aloft a threatening slaughter knife while demanding that he repays his debt to her. Also, there are interpolated shots of the slaughter of cattle with which he has been symbolically associated from the start of the film. Haunted in this way, he is unable to board and is shown fleeing from the ship. Anta apparently continues with the journey although she immediately experiences racist remarks from the white passengers and, in keeping with the elliptical narrative of the film, she is found, a little later, still with Mori, on the beach, gazing out to sea.

The sense of travel from a pre-industrial hinterland to the metropolitan centre is thus presented more in the form of phantasmagoria, from Mory's haunting memories to the siren-like invitation of songs about Paris on the soundtrack and the fantasy of a triumphant return surrounded by luxury. Where the journey in *Blood of the Condor* carefully underlines the physical distances which separate incompatible outlooks upon the world, the symbolic role it plays in *Touki-Bouki* is associated with the psychological stresses arising from this sense of incompatibility. In the everyday shanty town all kinds of exchange between worlds are shown to be readily progressing, emphasised by the noise of a jet aircraft overhead or the plodding postman who laboriously carries a letter from a far distant place.

Cultural difference is understood here in the circumstances of the exigency of shanty town life where exchange must take place despite the psychological stresses. Rather than trying to claim the binaries of tradition and neo-colonialist modernity as irreducible and observable opposites these binaries become extrapolations of the characters' mindset.

Central to this mindset is the combination of the material lure of economic migration with a wider sense of wanderlust, sustained by the rhetoric of tourism. Instead of being shunned as politically incompatible with the true interests of indigenous peoples, as in *Blood of the Condor*, the notion of tourist travel becomes a means for exploring the grip of neo-colonialist ideology. Promotion of the impulse to tourism is suggested in a conventional way by sets of images in songs and travel posters extolling the wonders of Paris. Yet the impulse to tourism is not the prerogative of a class, as we have seen it represented in *Blood of the Condor*, but a matter of individual consumption. This is accompanied by an absence of explicit political discourse in *Touki-Bouki*. Instead, the film is preoccupied with conflicts between collective identity, on the one hand, and, on the other, the individualism and consumerism that underlie Anta and Mory's desire to travel to France. Collective identity is evoked near the beginning of the film as Anta and Mory ride their bike through the shanty town. At first people are shown dodging the bike but a group of children begin to run alongside, keeping pace with both riders and the camera. These scenes presage fractious relationships around the sense of the collective, yet without any specific political articulation. There is, for instance, a comic fight between women at the water tap while the crowd cheers them on. Later, Anta and Mory try to steal the takings from a wrestling match. While the crowd participation and celebration of the victorious contestant in the stadium manifest an exuberant vigour, the couple hunt around at the back of the stand for the money. But the purloined trunk, which they suppose to contain the takings, proves only to be filled with traditional fetishes, indicating a set of non-monetary values that cannot be assimilated to their purpose.

These encounters with collective identity, then, can only be experienced as series of disconnected, comic or bizarre encounters, signalling a decaying and socially fragmented world. As an elliptical narrative *Touki-Bouki* is closer to modernist cinema yet it avoids nihilism through the energy of these fragmented representations loosely linked together as episodic encounters of a travel narrative. This energy also makes it hard to tie down the meaning of these collective symbols or to place limits upon their implications. In particular, despite its fragmentation, collective identity cannot be simply assigned to the past in contrast with an individualist future. The coincidence of the travel narrative with linear notions of social progress – progress away from a collective traditional past into an individualist modernity – is thus thrown into doubt.

This uncertainty generated about what kind of political position to take up in relation to the film's battle of symbols counters the kind of Orientalist fixing of positions on which Chow comments. The metropolitan viewer, outside the immediate African context, is left uncertain where to stand in the light of the film's satiric intent.6 Is it hostile to collective interest or to individual desires? Anta and Mory's journey through Senegalese life, propelled by their desire to travel abroad, generates collisions with an oppressive collective identity. Yet, at the same time, it is the symbols of collective identity, despite lacking a political articulation, which bring into question their individualist pursuit of travel.

Mutual critique between tourism and politics

As we have seen with *Baraka*, the communal aspect of pre-industrial life is emphasised for the virtual tourist gaze as an alternative to the despoliation of modernity. By contrast, in *Touki-Bouki* the collective viewpoint of tradition is threatening and haunting rather than a spectacle for the tourist. It is these kinds of dynamic relationships, arising from the representation of journeying in the Third World, that come into play in a number of subsequent films, even though traditional community life in these cases is generally open to appropriation by the tourist gaze in a way that it is not in *Touki-Bouki*. In the same vein as *Touki-Bouki*, however, none of these films can be construed as representing fixed and unalterable relationships between two worlds to be simply recorded as reality for the contemplation of the distant viewer.

For instance, in Sanjines' *The Secret Nation* (Bolivia, 1989), made twenty years after *Blood of the Condor*, an area of the Bolivian *altiplano* is represented, over a long historical time-span, as subject to social turmoil and cultural decay through economic pressures and conflict with the military. Tradition is slowly disappearing. Although the villagers are impoverished, radio and formal education are shown to be present while many are drawn into employment in the mines. Resistance to state power is represented through an alliance between the peasants and the miners who are involved in strikes and demonstrations.

Separate from this political conflict, the main character, Sebastian Maimani, is shown progressing across mountain and plateau, returning to his village after many years in the city. He walks home, carrying a large traditional mask and headdress, to expiate his betrayal of the community by stealing aid money that he had been entrusted to claim from the authorities. His expiation takes the form of the revival of the almost forgotten custom of a ritual dance to death. This is his way of acknowledging the continuing importance of the collectivity of the village from which he has become separated. While the rest of the village becomes engaged in a fully modern political struggle, Sebastian's journey to death suggests the importance of the cultural space he has abandoned, starting with his own alienation from traditional culture. His experience of alienation, shown in flashback, began when, as a child, he was given as a servant to an elite family in the city. In the city Sebastian has attempted to deny his own name and Aymara identity. The corruption of the city leads him to side with the military against the peasants by joining the army – before his ultimate role of betraying his village through theft. Yet, his journey to perform the ancient custom of dancing to death provides the cue to recall the historical events of his life-time. These are intertwined with his personal memories in the series of flashbacks during his long walk home.[7]

The film, then, contrasts different viewpoints upon the nature of collective identity. Sebastian's cultural memory is the motivation for the retrospective historical viewpoint of the experiences of the community. Yet it is not so clearly aligned with the political struggles in which the villagers are engaged in the present. His alienation from his culture has made him prey to the individualism of the city but he is the one still moved by the significance of the dance to death. His performance of the dance coincides with the return of the villagers from a violent clash with the authorities in support of the miners. They upbraid him for dancing in such circumstances. Yet his longer memory of the old ritual of dancing to death recalls the traditional fate of exclusion from collectivity in the past. The prolonged confrontation as the marchers pass by the scene of his dance indicates a problem of access to the self-contained space of collective identity. Is it in the solidarity of the marchers with the cause of the striking miners? But these marchers are, with the exception of the very old, no longer acquainted with the tradition of the dance. Or is it with Sebastian who knows what he has lost by being condemned to individualistic exclusion? Sebastian cannot be accommodated within the organised resistance to defend the community offered by his militant brother Vicenze – that is through subscription to mass struggle – and is therefore excluded along with his attachment to the pure cultural memory of the community.

From the metropolitan viewer's standpoint Sebastian's rediscovery of a pure allegiance to traditional culture offers the prospect of absorption within the otherness of Aymara culture, which calls him back to his roots. Yet the dance to death is set up as an incongruous confrontation with the marchers on a narrow hill path. They are carrying their dead and wounded while displaying their own brightly coloured banners, fluttering in the breeze as a contrast to Sebastian's traditional headdress. Angry incomprehension of Sebastian's ritual among those returning from a bloody confrontation in support of the striking miners is played against the haunting evocation of a 'disappearing' culture. Sebastian, in traditional costume and headdress is filmed from a low angle, dancing to death against the blue sky and the mountains beyond. What might seem to have most appeal to the tourist gaze is anathema, though, in the eyes of the strikers. The logic of this desire for pure tradition becomes apparent at the end of the film as Sebastian is shown watching his own funeral from a distance. In this case the vanishing point of pure tradition sought out by Sebastian, as he attempts to escape his alienated position in modernity, is clearly situated beyond the world in which the other characters move.

The travelling relationship between the city and the hinterland then sets up the possibility of mutually critical perspectives for the metropolitan viewer between virtual tourism and political understanding. This film moves beyond earlier Sanjines/Grupo Ukamau films such as *The Courage of the People* (Sanjines, Bolivia, 1971) which sums up the whole experience of a village in terms of a strike against the mine owners and a massacre by the army. Here the metropolitan superiority in judging the Third World politics is challenged by the ambivalence of the conflict between Sebastian's attachment to a forgotten tradition and the urgency of contemporary struggles. It is just not possible to assay the relative political weight of these two perspectives from the metropolitan point of view. On the one hand, we are given a sympathetic insight into Sebastian's political detachment from the rest of the village, reinforced by the otherness of his ritual as a spectacle for the tourist gaze. On the other hand, Sebastian's and therefore the virtual tourist's investment in recreating the ancient tradition of dancing to death is brought into question as a politically negative act – in contrast to the fêting of spiritual collectivity in such films as *Baraka*.

The metropolitan experience of virtual travelling culture

In the last two Chinese examples to be examined here the notion of the peasant community has already been subject to a massive degree of ideological investment in China's progress to modernity. Mao's communist movement was substantially built around a theory of revolution based upon the peasants, promoted from his remote rural headquarters in Yanan, a location which underlined his claim to identify with the rural mass of China's population. From the beginning of the post-Mao 'Fifth Generation' of cinema in the middle of the 1980s, a strand of film making, exemplified by *Yellow Earth* (Chen Kaige, 1984), ambivalently re-examines the role of peasant life in relation to the modern state. The peasant community, supposed to be central to Mao's revolution, is the starting point in many of these films for journeys which test the effectiveness of the contemporary bureaucracy and the increasingly free market economy. What is most significant for the metropolitan spectator here in following the narratives of these journeys, is the problematic orientation towards the peasant community as an entity on the margins of modernity.

The Story of Qiu Ju (Yimou, China/Hong Kong, 1992), shot with multiple cameras to give a degree of documentary immediacy, concerns the pregnant Qiu Ju. She visits increasingly elevated officials in a series of cities to argue her husband's case for compensation after he has been assaulted by the village headman. In the process she encounters an ever more perplexing modernity. Yet, after each city visit she returns to a village by a snow-covered road, placed by a long-shot as a remote self-contained entity in a steep valley and endowed with a touristic gloss through the bright colours of the peppers hung out to dry on the front of the houses. In working through questions of justice for peasants in contemporary China the metropolitan spectator is placed in the kind of Orientalist mode described by Chow, making judgements on the inefficient and remote workings of justice and bureaucracy in a highly centralised state. Yet there is a significant ambivalence about the connection between peasant communities and contemporary Chinese life. The persistence of Qiu Ju in taking her complaint to increasingly higher officials and larger cities is finally given a response and the village headman is arrested after an X-ray reveals that her husband's ribs were broken during the original assault. However, by the time the law has deliberated on the case, the headman has been instrumental in getting an ambulance for Qiu Ju when her life is threatened in giving birth. The law intervenes at a point when the strength of communal ties of the village has come to seem more important than official judicial decision. The whole political and judicial process appears thus to be detached from village life.

In Zhou Xiaowen's *Ermo* (China, 1994) Ermo works through the night to make noodles in order to acquire the largest television set in the county and thus settle a score with her neighbour who already has a set. She represents an ironic Stakhanovite entry into the market economy of modern life, represented by the distant town to which she laboriously journeys in order to sell her wares. At the end of the film, the television set, the fruit of Ermo's obsessive individualism, is squeezed into her family's single room home. Yet, collective identity reasserts itself as a packed assembly of attentive and curious villagers gather there to watch while Ermo falls into exhausted sleep. This juxtaposition of collectivism and individualism is central to the film's interrogation of the notion of progress. The peasants' comments as they watch television induce a sense of defamiliarisation from what, for the metropolitan viewer, has become a casual element of the domestic environment. The equivocal stance of both the film maker Zhou and the metropolitan viewer is indicated by Tony Rayns' defence of Ermo against a metropolitan position of knowledge:

> ... although Ermo's naiveté generates much of the film's humour, Zhou never tries to get laughs at her expense. When Ermo first sees the television set in the town store, for example, it's playing a Chinese dubbed tape of a western softcore sex film to a silent rapt audience of peasants. Ermo registers her bafflement that the foreigners are speaking Chinese, but the real joke is far more complex. What does it mean to show a shoddy foreign sex movie to people who have never met a foreigner or, indeed, travelled any further than the nearest county town?
>
> (Rayns 1995: 48)

In both films, then, the way in which apparently traditional communities are connected to contemporary Chinese modernity is at issue, emphasised by the endless and tortuous journeys to distant towns. For the metropolitan spectator this becomes an issue of the perception of temporality. Fredric Jameson has argued:

The modern still had something to do with the arrogance of city people over against the provincials, whether this was a provinciality of peasants, other and colonised cultures, or simply the pre-capitalist past itself: that deeper satisfaction of being '*absolument moderne*' is dissipated when modern technologies are everywhere, there are no longer any provinces, and even the past comes to seem like an alternate world, rather than an imperfect, privative stage for this one.

(Jameson 1994: 11)

This uncertainty about where the margins of modernity lie relates to his argument that the cinema as an invention of modernity conforms to a kind of modularity, of standardisation of change and of social life:

where intensified change is enabled by standardisation itself, where prefabricated modules, everywhere from the media to a henceforth standardised private life, from commodified nature to uniformity of equipment, allow miraculous rebuildings to succeed each other at will.

(Jameson 1994: 16)

The effects of this standardisation also lie at the heart of Rey Chow's argument about the processes of cultural exchange at work in the emergence of 'fifth generation' Chinese cinema during the 1980s. Drawing on the film theory of Pasolini, she emphasises the view that the 'mode of time' of the cinema is 'belated and retrospective, it is also concurrent and forward-looking' (Chow 1995: 42). In terms of attitudes to modernity 'it is perhaps only in film that the ambivalence characteristic of modernity – a demand for a brand new beginning that is at the same time an intense look back to the past – becomes fully materialised' (Chow 1995: 41). The 'brand new beginning' here is the reconstruction of ideas of China in the aftermath of the traumas of the Cultural Revolution. Yet the notion of cinema as central to 'miraculous rebuilding' of the Third World nation through a concurrence of retrospection and forward movement has also been at the heart of debates about Third Cinema in the aftermath of decolonisation. The question here for the metropolitan viewer is the definition of the starting point from which anticipation begins, one that can fix the relation to the margins of modernity in circumstances where definition of alternatives to modernity have become, as Jameson indicates, increasingly uncertain.

Stephanie Donald points out that in recent Chinese cinema 'Tradition is conflated with ideals of communist organisation, such as the rural idylls of collective farming' (1995: 338). In either film discussed here it is difficult to determine whether the villages are presented as traditional or already having passed through a modernising mass socialist movement. The ambivalence of the retrospection/anticipation couplet is stressed when the vanishing point of retrospection becomes uncertain in this way. In *The Story of Qiu Ju* the disjunction between the contemporary feel of the documentary style and the visual plenitude of otherness of the 'traditional' village offered to the spectator indicates the problem of allocation of this setting to a chronological period. As Donald puts it concerning recent Chinese cinema:

> the meaning of contemporary is not a chronological issue, but relies on the synchronicity of a popular consciousness at a particular time and in a particular place. Being contemporary is not a state that we can always assume to be our own, nor one that we necessarily share with those around us.
>
> (ibid.)

What the metropolitan viewer is confronted with is a problem of contemporary synchronicity while lacking an anchoring in popular consciousness of the kind of dynamic changes affecting these distant areas. If the Orientalist position on the political projects of the Third World is undermined by such uncertainty, the tourist component of involvement with the pre-modern is caught up in a dynamic exploration of time/space relationships which brings its own position into question. The tourist gaze becomes caught up with problems in defining its own synchronicity and is therefore no longer able to reinforce its vantage points by a sense of certainty about its position in historical progress. It is not necessary, though, in considering the metropolitan spectator's encounter with these films, to see this as a case of a modernist era being replaced by a postmodernist dominant, according to Jameson's theories. Rather, these films are a starting point for recognition that the historical sense of progress associated with the notion of modernity may be subject to the renegotiations of travelling cultures.

Notes

1 Contrary to my position, Mike Wayne has argued that Third Cinema continues to be a relevant category. However, his examples tend to work through allegory or other more indirect ways of reflecting on neo-colonialism or First World hegemony rather than the directly militant approach of the late 1960s (Wayne 2001).
2 The term 'Fifth Generation' refers to a particular cohort passing through the Beijing Film Academy. It should be noted that the recent films of Zhang Yimou, such as *Not one Less* (1998) and *The Road Home* (1999), do not present pre-industrial peasant life as a world apart as signalled by pictorial aetheticisation.
3 The notion here of mutually critical relationships between discourses arises fromconsidering films as novelistic texts, that is they are composed of a multitude of competing discourses (Bakhtin 1981). Bakhtin has been argued to consider novelisation as providing the dominant aesthetic form of modernity (Hirschkop 1999). For a specific consideration of novelisation in relation to film and television see John Caughie (1991).
4 'The colonised man who writes for his people ought to use the past with the intention of opening up the future, as an invitation to action and a basis for hope' (Fanon, 1967: 187). On the other hand, Fanon sees no future in preserving traditional culture in a 'mummified' form.

5   Sanjines founded the Grupo Ukamau with his friends Oscar Soria and Ricardo
    Rada. 'Ukamau' means 'That's the way it is' in Quechua (Hess 1993).
6   This uncertainty is noted by Nwachukwu Frank Ukadike who suggests that
    Europeans would tend to see the film as an 'avant-gardist manipulation of
    reality' while in the African context it would be read as an indictment of contem-
    porary African society (Ukadike 1994: 173).
7   For an account of the historical context see King (1990: 197).

Bibliography

Bakhtin, M.M. (1981) *The Dialogic Imagination*, trans. Caryl Emerson and Michael
    Holquist. Austin, TX: University of Texas Press.
Caughie, J. (1991) 'Adorno's Reproach: Repetition, Difference and Television Genre',
    Screen, 32 (2): 127–153.
Chow, R. (1995) *Primitive Passions – Visuality. Sexuality. Ethnography and Contem-
    porary Chinese Cinema*, New York: Columbia University Press.
Clifford, J. (1992) 'Traveling Cultures', in Lawrence Grossberg, Cary Nelson and
    Paula Treichler (eds) *Cultural Studies*, London: Routledge, pp. 96–112.
Donald, S. 'Women Reading Chinese Films: Between Orientalism and Silence',
    Screen, 36 (4): 325–340.
Fanon, F. (1967) *The Wretched of the Earth*, Harmondsworth: Penguin.
Hess, J. (1993) 'Neo-realism and Latin American Cinema', in John King, Ana M.
    Lopez and Manuel Alvarado (eds) *Mediating Two Worlds – Cinematic Encounters
    in the Americas,* London: British Film Institute, pp. 104–118.
Hirschkop, K. (1999) *Mikhael Bakhtin – An Aesthetic for Democracy*, Oxford:
    Oxford University Press.
Jameson, F. (1994), *The Seeds of Time*, New York: Columbia University Press.
King, J. (1990) *Magical Reels – A History of Cinema in Latin America*, London:
    Verso.
Loshitzky, Y. (1996) 'Travelling culture, travelling television', *Screen*, 34 (4): 323–335.
Rayns, T. (1995) Review of Ermo, *Sight and Sound*, 5: 7, 48.
Roberts, Martin (1998) '*Baraka*: World Cinema and the Global Culture Industry',
    Cinema Journal, 37 (3): 62–81.
Solanas, F. and Getino, O. (1983) 'Towards a Third Cinema', in Michael Chanan
    (ed.) *Twenty Five Years of the New Latin American Cinema*, London:
    BFI/Channel Four, pp.17–27.
Stam, R. (1987) 'The Hour of the Furnaces and the Two Avant-Gardes', in Coco
    Fusco (ed.) *Reviewing Histories: Selections from New Latin American Cinema*,
    Hallwalls: Buffalo, pp. 90–105.
Wayne, M. (2001) *Political Film – The Dialectics of Third Cinema*, London: Pluto
    Press.
Ukadike, N. F. (1994) *African Cinema,* Berkeley: University of California Press.
Urry, J. (1990) *The Tourist Gaze –- Leisure and Travel in Contemporary Societies*,
    London: Sage.

Index

eBooks

eBooks – at www.eBookstore.tandf.co.uk

A library at your fingertips!

eBooks are electronic versions of printed books. You can store them on your PC/laptop or browse them online.

They have advantages for anyone needing rapid access to a wide variety of published, copyright information.

eBooks can help your research by enabling you to bookmark chapters, annotate text and use instant searches to find specific words or phrases. Several eBook files would fit on even a small laptop or PDA.

NEW: Save money by eSubscribing: cheap, online access to any eBook for as long as you need it.

Annual subscription packages

We now offer special low-cost bulk subscriptions to packages of eBooks in certain subject areas. These are available to libraries or to individuals.

For more information please contact webmaster.ebooks@tandf.co.uk

We're continually developing the eBook concept, so keep up to date by visiting the website.

www.eBookstore.tandf.co.uk